MAMMAL BONES AND TEETH
AN INTRODUCTORY GUIDE TO METHODS OF IDENTIFICATION

MAMMAL BONES AND TEETH

An Introductory Guide to Methods of Identification

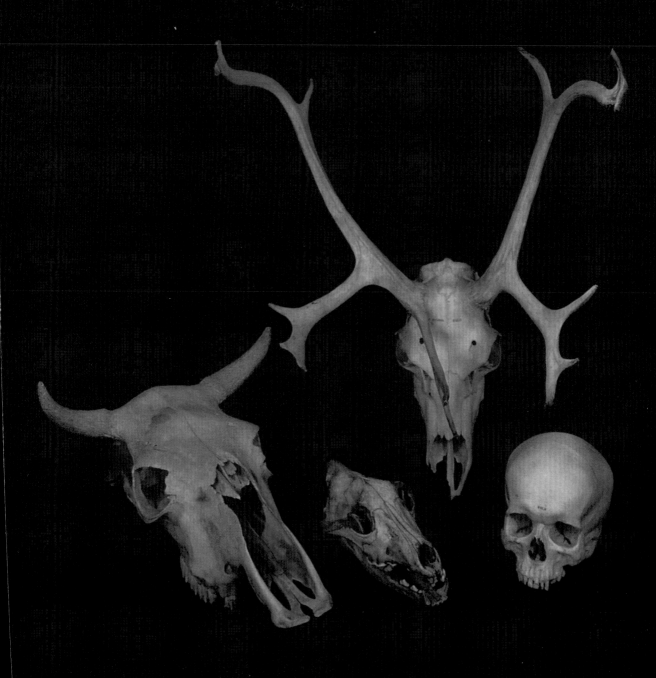

SIMON HILLSON

MAMMAL BONES AND TEETH

AN INTRODUCTORY GUIDE TO METHODS OF IDENTIFICATION

Simon Hillson

Routledge
Taylor & Francis Group

LONDON AND NEW YORK

Originally published by University College London Institute of Archaeology

First published 2009 by Left Coast Press, Inc.

Published 2016 by Routledge
2 Park Square, Milton Park, Abingdon, Oxon OX14 4RN
711 Third Avenue, New York, NY 10017, USA

Routledge is an imprint of the Taylor & Francis Group, an informa business

Library of Congress Cataloging-in-Publication Data available from the publisher.

ISBN 978-0-905853-30-7 paperback

CONTENTS

LIST OF FIGURES

ACKNOWLEDGEMENTS

First of all I would like to thank my wife Kate and sons William and James who as with previous projects have put up with a great deal. I am also grateful to more than ten generations of students who, although they may not have known it at the time, tested most of my ideas in previous versions of this guide. In addition, I have benefited from the comments of others who used earlier versions when they were teaching themselves the business of identification. Several colleagues – James Rackham, Don Brothwell and Dale Serjeantson – very kindly read final draughts and made suggestions. I take, of course, sole responsibility for any errors or idiosyncrasies, especially as I was not always able to follow their advice in full! Most of the reference material came from the collections at the Institute of Archaeology, University College London, with the addition of some specimens of my own. I must also thank Dr. Juliet Clutton-Brock of the Natural History Museum, London, who gave me access to goat and bison material. The process of publication was co-ordinated by Jim Black of Archetype Books, and I am grateful for his energy and encouragement throughout.

INTRODUCTION AND AIMS

This book has developed out of literature for undergraduate courses in zooarchaeology, first at Lancaster University and then at the Institute of Archaeology, University College London. It is designed very much as an introduction and its aim is to help students achieve an initial level of basic knowledge, from which they can expand later. It is intended to be used in conjunction with a reference collection of bones and teeth, and with tuition from a zooarchaeologist. Much work with fragmentary archaeological remains involves recognising small differences in shape and to achieve this, students need to spend many hours handling reference specimens on their own, ideally making their own drawings and notes. This book is intended to help at this early stage by highlighting the main points on which identifications are made. It is not intended to be used on its own without named reference material.

ANIMALS INCLUDED

It is best to start with just a few species and, fortunately, the bulk of the most common large mammal remains on Eurasian and North American sites can be fitted into a compact group. This includes the most widely ranging domestic animals, together with their wild relatives and ancestors, and the wild deer which are found throughout Europe, northern Asia and North America. Domestic animals are a taxonomic minefield, and an arbitrary decision has been taken to follow Corbet & Hill (1986) here:

- **Horse** *Equus ferus* Boddaert, 1785 – wild and domestic forms.
- **Cattle** *Bos primigenius* Bojanus, 1827 – wild and domestic forms (domestic cattle are often differentiated by other authors as *Bos taurus*).
- **Bison**, American Buffalo or Wisent *Bison bison* Linnaeus, 1758 – both American and European forms are here included in one species, but other authors often refer to European bison as *Bison bonasus*, reserving the term *Bison bison* for the American Buffalo.

- **Sheep** *Ovis ammon* Linnaeus, 1758 – wild and domestic forms (domestic sheep are often distinguished elsewhere as *Ovis aries*).
- **Goat** *Capra aegagrus* Erxleben, 1777 – wild and domestic forms (domestic goat are often distinguished by other authors as *Capra hircus*).
- **Moose** or European Elk *Alces alces* Linnaeus, 1758.
- **Caribou** or Reindeer *Rangifer tarandus* Linnaeus, 1758.
- **Red Deer**, Wapiti or North American Elk *Cervus elaphus* Linnaeus, 1758.
- **Fallow Deer** *Cervus dama* Linnaeus, 1758 (distinguished by many other authors as *Dama dama*).
- **Roe Deer** *Capreolus capreolus* Linnaeus, 1758.
- **Pig** *Sus scrofa* Linnaeus, 1758 – wild and domestic forms.
- **Dog** or Wolf *Canis lupus* Linnaeus, 1758 – wild and domestic forms (domestic dogs are frequently referred to as *Canis familiaris* by other authors).
- **Cat** *Felis silvestris* Schreber, 1777 – wild and domestic forms (domestic cat is often called *Felis catus* in the literature).
- **Human** *Homo sapiens* Linnaeus, 1758.

In the text below, the first (bold) name is used.

Cattle, bison, sheep and goat are all anatomically rather similar and, where they are considered all together, their common family name is used – **Bovidae** or **bovids** for short. Cattle and bison are not only very alike in many aspects of the teeth and skeleton, but are both of a similarly large size and robust build and are therefore grouped together as **large bovids** throughout most of the text. In the same way, sheep and goat are very close anatomically, but are smaller in size and slighter in build and can be grouped as **small bovids** for the purposes of this book. It should be pointed out that there are many other bovids, small and large, in the world which are not considered here. Even so, the family as a whole shows relatively little variation throughout the skeleton, except for an exotic variety of horns, and many of the features described here are equally characteristic of other bovids. Distinguishing between cattle and bison, sheep and goat, is a specialist job which is largely beyond the scope of this introductory book, although it is routinely carried out in zoo-archaeological reports. The skull and the bones of the feet provide the simplest distinctions within the large bovid and small bovid categories, and are the only ones to be described in this book. Further details for sheep/goat distinction are given in Boessneck *et al.* (1964), Boessneck (1969), Payne (1969; 1985), Prummel & Frisch (1986) and Clutton-Brock *et al.* (1990). Differences between cattle and American Buffalo are given by Olsen (1960), McCuaig Balkwill & Cumbaa (1992), and Reynolds (1939) is still the best discussion of the differences between wild cattle and the giant fossil bison of Europe.

Similarly, moose, caribou, red deer, fallow deer and roe deer are anatomically alike and are frequently referred to by their shared family name – **Cervidae** or **cervids** for short. There are differences in size. Moose is by far the largest, followed by red deer. Caribou and fallow deer are intermediate in size and roe deer markedly smaller. It is therefore convenient to summarise them as **large**, **intermediate** and **small** cervids. The differences between species are again most apparent in the head region (the antlers) and in the feet, and it is only for these parts of the skeleton that detailed distinguishing features are described here.

Although the descriptions and figures apply specifically to the animals listed above, they include many features which are characteristic of horses (Equidae), bovids, cervids, cats (Felidae), dogs and their relatives (Canidae) and Primates as a whole. It is a only short step from these basic distinctions to a wider range of identifications.

TERMS, ABBREVIATIONS, DIVISIONS AND DIRECTIONS IN THE SKELETON AND DENTITION

Non-perishable remains of mammals are divided into the skeleton (bones) and dentition (teeth). Each has its own system of names.

THE SKELETON

In this book, bones are labelled for all animals as though they are quadrupeds. This is not the normal practice in human anatomy, but is done here to minimise the number of terms used. Using quadrupedal terminology assumes that the feet and palms are flat on the floor, with the thumbs innermost, and that the head is pulled right back so the nose points straight forwards in the direction of travel (Fig. 1). This is a relatively uncommon position for a human being to assume (not to mention an uncomfortable one), and makes a great deal of difference to the terminology of the forelimb (below).

One key concept in the system is the *median sagittal plane*. This is the imaginary plane which would divide the skeleton into two equal, mirror image, left and right halves. Any one bone in the skeleton has six surfaces, or orientations in which it can be viewed. The names for four of these surfaces are common throughout most of the skeleton:

Medial – facing towards the median sagittal plane.

Lateral – facing away from the median sagittal plane.

Cranial – facing towards the front of the skull.

Caudal – facing towards the tip of the tail.

Axial skeleton

The axial skeleton comprises the *skull*, the *vertebral column*, the *ribs* and the *sternum* or breastbone. Two additional names apply to this part of the skeleton:

Dorsal – facing towards the back.

Ventral – facing towards the belly.

Appendicular skeleton

This consists of the two pairs of limbs, the more cranially placed are here called the left and right *forelimbs*, and the more caudally placed, the left and right *hindlimbs*. Each forelimb is attached to the axial skeleton at the sternum by the *clavicle* or collar bone. This is joined at the shoulder to the *scapula* (shoulder blade), which is held in position against the dorsal part of the ribcage by muscles. Also joining the scapula at the shoulder is the *humerus*, which in turn joins the paired *radius* and *ulna* at the elbow. The extremity of the forelimb is called the *manus* (*carpals*, *metacarpals* and manus *phalanges*) or wrist and hand. Each hindlimb is attached to the axial skeleton by one of the *innominate bones* which, together with the base of the vertebral column, form the *pelvic girdle*. This is connected at the hip joint to the *femur*, which in turn joins the paired *tibia* and *fibula* at the knee. The extremity of the hindlimb is called the *pes*

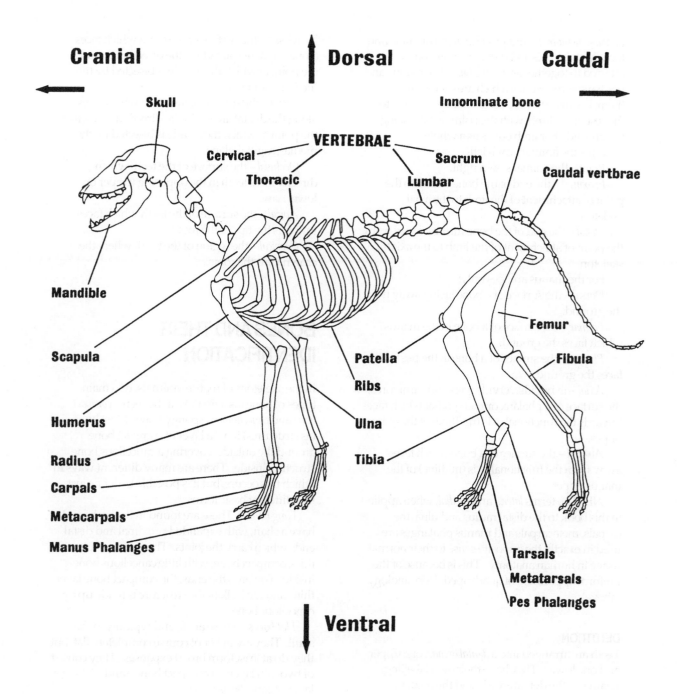

Cranial

Dorsal

Caudal

Skull

Cervical

VERTEBRAE

Innominate bone

Sacrum

Thoracic

Lumbar

Caudal vertbrae

Mandible

Scapula

Patella

Ribs

Femur

Fibula

Humerus

Ulna

Radius

Tibia

Carpals

Metacarpals

Manus Phalanges

Tarsals

Metatarsals

Pes Phalanges

Ventral

FIGURE 1. THE SKELETON OF THE DOG

3

(*tarsals, metatarsals* and pes *phalanges*) or ankle and foot. Metacarpals and metatarsals are often referred to together as *metapodials*. The manus and pes have some names which are exclusive to them, because they are set at a different angle to the rest of the limb, with the palm or sole facing the ground. For limb bones as a whole:

Left – the animal's own left.

Right – the animal's own right.

Proximal – the end of the bone closest to the point of attachment of the limb to the axial skeleton.

Distal – the end of the bone furthest away from the point of attachment of the limb to the axial skeleton.

For the manus and pes only:

Dorsal – the surface of a bone facing away from the ground.

Palmar – the surface of a bone in the manus which faces the ground.

Plantar – the surface of a bone in the pes which faces the ground.

Axial – in bovids, cervids, pigs and carnivores, the surface of a phalanx or metapodial which faces towards the functional axis (midline) of the manus or pes.

Abaxial – the surface of a bone which faces away from the functional axis (midline) of the manus or pes.

NB. The terms *lateral* and *medial*, when applied in this book to the distal radius and ulna, the carpals, metacarpals and manus phalanges are used in exactly the opposite sense to their normal usage in human anatomy. This is because of the uniform application of quadrupedal terminology (above).

DENTITION

Teeth are arranged into a *dental arcade*, one upper and one lower. They have their own labelling, relative to the dental arcade and the median sagittal plane, which splits both arcades into two equal and opposite halves. In this book, the following labels are used:

Vestibular – the surface of the tooth facing the lips and cheeks, outside the dental arcade. This term is equivalent to *labial* or *buccal*, used in human dental anatomy.

Lingual – the surface of the tooth facing the tongue, inside the dental arcade. The term *palatal* is also used with this meaning in human dental anatomy.

Mesial – the surface of the tooth which faces along the dental arcade in the direction towards the point at which the arcade is bisected by the median sagittal plane.

Distal – the surface of the tooth which faces along the dental arcade in the direction away from the point at which the arcade is bisected by the median sagittal plane.

Occlusal – the surface of the teeth which directly faces teeth in the opposing (upper or lower) jaw.

Apical – the surface of the teeth which faces towards the tips of the roots.

Cervical – the region of the tooth where the crown joins the roots.

BONES AND THEIR IDENTIFICATION

In the skeletons of mature animals, two main types of bone (as a tissue) can be seen – *compact bone* and *cancellous* or spongy bone. Most bones (as structures) have a layer of compact bone around the outside, covering a cancellous bone structure inside. There are many different ways in which this occurs, but it is possible to define just five different basic forms:

Long bones – These are found in the limbs and have a shaft, with expanded proximal and distal ends which carry the joints. The shaft is a tube of thick compact bone, with little cancellous bone inside. Towards the ends, the compact bone layer thins and the bulk of the structure is made up by cancellous bone.

Flat bones – These are found typically in the skull. They are never of course completely flat, but they do at least form broad expanses. They consist of two thin layers of compact bone, sandwiching a layer of cancellous bone.

Cubic bones – Typically, these are the small carpal and tarsal bones. They are never precisely cubic, but the term does convey their relative regularity. They are largely composed of cancellous bone, with a thin coating of compact bone.

Irregular bones – Vertebrae are the usual example of these. They have a complex form, most of which is composed of cancellous bone, but the surface coat of compact bone varies considerably in thickness and, in parts, most of the structure may be of compact bone.

Sesamoid bones – Usually small bones, embedded in tendons where they pass over a joint. There are many in the manus and pes, and the largest is the patella or kneecap.

All the various bones of the body fit against one another through a limited number of types of joints or *articulations*. The commonest is the *synovial joint*, where the bearing surfaces (usually called *facets*) are of compact bone, overlaid with a coating of cartilage. The flat bones of the skull are joined by *sutures*, bound closely together by ligaments. A similar arrangement (*gomphosis*) binds the teeth into their sockets. Some bones are joined together rather more flexibly by ligaments (*syndesmosis*) and the best example of this is the interosseous ligament linking radius and ulna in humans (below). In many growing bones, the actual growth takes place in cartilage which is subsequently replaced by bone itself. A cartilage growth plate separates different parts of the bone – for example a growing long bone is split into three elements, the main part including shaft (*diaphysis*), proximal and distal portions (*epiphyses*). The cartilagenous joint between these elements is called a *synchondrosis*. As the skeleton approaches maturity, these synchondroses are replaced by bone, and the epiphyses fuse on to the diaphysis. One final type of joint, the *symphysis*, is a combination of cartilagenous and ligamentous connections. The joint surfaces of the bones are again coated with a layer of cartilage, but these are bound together with ligaments. The classic example of this is the pubic symphysis, joining the two innominate bones together at their ventral contact. Another is the "disc" joints between the bodies of vertebrae. Symphyses also change in structure with age and in some animals, for example, the pubic symphysis is eventually replaced by a bony connection.

Several general terms are used for features of individual bones:

Body – the main part of the structure.

Process – a prominence, clearly extending out from the body of the bone.

Spine – a pointed process.

Tuberosity – a rounded, usually roughened, bulging feature, for the attachment of ligaments and muscles.

Tubercle – usually a smaller tuberosity.

Fossa – a depression or pit (plural *fossae*).

Articulation / articular surfaces – the elements of the bone which make a joint with a neighbouring bone, usuallt defined by a clear border.

Facet – a small, usually flat articular surface.

Head – a rounded, convex articular surface.

Glenoid cavity or *acetabulum* – a dished or concave articular surface.

Condyle – a cylindrical articular surface, or one which forms part of a cylindrical joint.

Trochlea – a pulley-like articular surface.

Foramen – a perforation for blood vessels or nerves (plural *foramina*).

Most archaeological remains are fragmentary and the descriptions below are split, not only into the different bones, but also into different elements of bones. There is not space to cover everything and it is not desirable to do so in an introduction, so only a selection of skeletal elements is included. Readers will not, for example, find details of the clavicle, sternum or patella even though they do occur on archaeological sites. Similarly, the diagrams are not designed to be completely faithful representations of particular specimens. Bones vary so much in one species that this would, in any case, not be very useful. The diagrams are best seen as cartoons, which emphasise the main distinguishing features and are there only to illustrate points in the text. Both diagrams and descriptions should be used in conjunction with a previously assembled reference collection. Another approach when learning is to use the archaeological collection under study as the reference collection and to make comparisons between specimens. Fine artwork of particular specimens, with detailed shading etc. can be found in Schmid (1972) and, at large scale, in Pales & Garcia (1981).

Identification of a bone or tooth specimen may be divided into four stages:

1. Decide which bone in the body it is from, and which fragment of that bone it is. This book is divided into sections dealing with the different bones and teeth. Each section has a short description of the features which distinguish a bone or tooth from others in the body. Quite a large proportion of specimens fail at this first stage of analysis, and must be counted as "unidentified".

5

2. Decide which side of the body it is from. This may be done by comparison with the figures or, in some cases, by particular features which are listed in the text.

3. Assess its size and robustness. Is it large (cattle or horse size), intermediate (human or large deer size), small (small deer, sheep, goat, pig, dog) or very small (cat).

4. Look for detailed points of anatomy to make any finer distinctions that are possible.

Some bones are more identifiable than others. So, for example, this book gives great detail on the metapodials, but rather less on the tibia. Similarly, some groups of animals, notably the bovids and cervids, show relatively little difference. The real answer in these cases is to assemble all available specimens of the bone in question on one table top and then to make comparisons between them. For example with sheep and goat metacarpals one may well end up with a series of specimens running from "clearly goat", through "could be sheep or goat", to "clearly sheep". This kind of direct comparison is highly recommended, particularly at the learning stage.

GENERAL NOTE ABOUT THE DRAWINGS

The figures show bones and teeth from the **LEFT SIDE** of the body. Each bone is a representation of the mature form, when growth has ceased. They have been drawn from specimens of medium size and many of the specimens with which they will be compared will be considerably larger or smaller. All are reproduced at **one half full-size**, with the exception of the teeth and the antlers. Each figure includes views in several orientations and these are defined by the "orientation boxes" drawn against each, labelled with the terms defined above. Clearly delimited articular surfaces are shaded. Many of the drawings include sections of bone shafts, which are shaded with a darker tone. Further details of the tooth drawings are given below.

SIZE AND VARIATION

Size and robustness are two of the main initial criteria for identification, particularly amongst the bovids and cervids, but they show considerable variation even within the same species. This may partly be due to the effects of domestication and subsequent breeding but, even in wild forms, there is variation between males and females, and between different populations of the same species. The red deer living today in Britain are considerably smaller than, for example, German red deer and also smaller than prehistoric British red deer would have been. The extinct giant bison of Eurasia was also considerably larger than the living European bison. It is usually given its own separate species name *Bison priscus* but, in all respects other than size, its bones are indistinguishable from those of the living forms. The aurochs or urus, extinct ancestor to domestic cattle, was of a similar build to the giant bison, very much larger and more robust than the earliest domestic cattle. Since that time, selective breeding has produced larger domestic cattle, the biggest bulls of which overlap in size and robustness with the smaller ancient wild cattle. The largest specimens of wild cattle represent enormous creatures, estimated to have been around 2.5m at the shoulder. Reference specimens prepared from modern cattle carcases are too small and lightly built to represent these, but they are also much too large to represent the majority of prehistoric domestic cattle, unless particularly small, rare breeds are chosen. There is not space in this booklet to represent the enormous range of variation in cattle size, and identification has instead to be carried out on the basis of proportions of bones and detailed features of their anatomy which are independent of size.

Pigs show a similar range of variation and prehistoric domestic forms are again markedly small, in relation to both wild boar and modern domestic breeds. Prehistoric horses, on the other hand, were about the same size as the wild form – both would be considered ponies in terms of the modern domestic breeds. The large horses bred for cavalry and to pull wagons and ploughs are one third again taller at the shoulder and have proportionately much more robust limb bones (indeed they were bred specifically for this

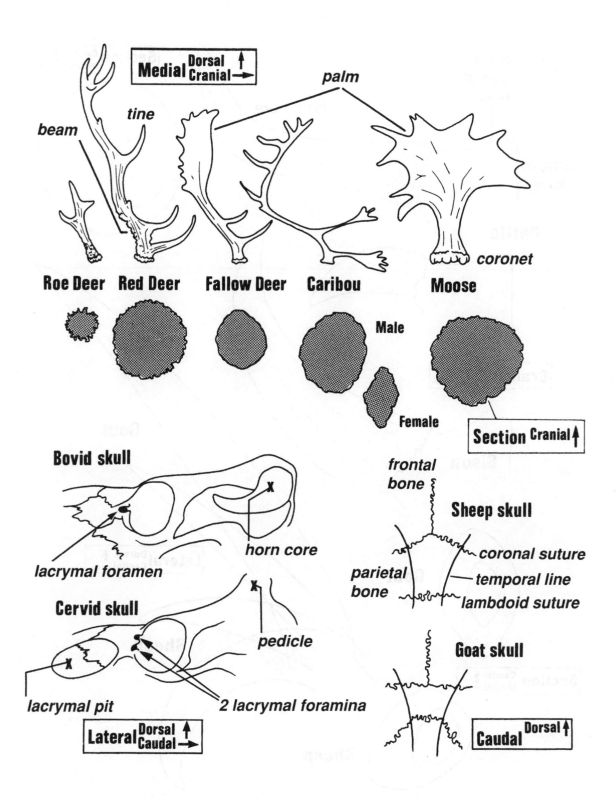

FIGURE 2. ANTLERS & SKULL FEATURES.

Top row: medial views of the left antler for roe, red deer, fallow, caribou and moose. Not to scale.

Middle row: sections of the antler beams, taken just above the brow tine, for roe, red deer, fallow deer, male and female caribou, and moose. Half life-size.

Lower left: lateral views of the region of the left orbit for a bovid (cattle) and a cervid (red deer), showing the orbital foramina and lacrymal pit. Not to scale.

Lower right: caudal-dorsal views of the coronal, sagittal and lambdoid sutures of sheep and goat. Not to scale.

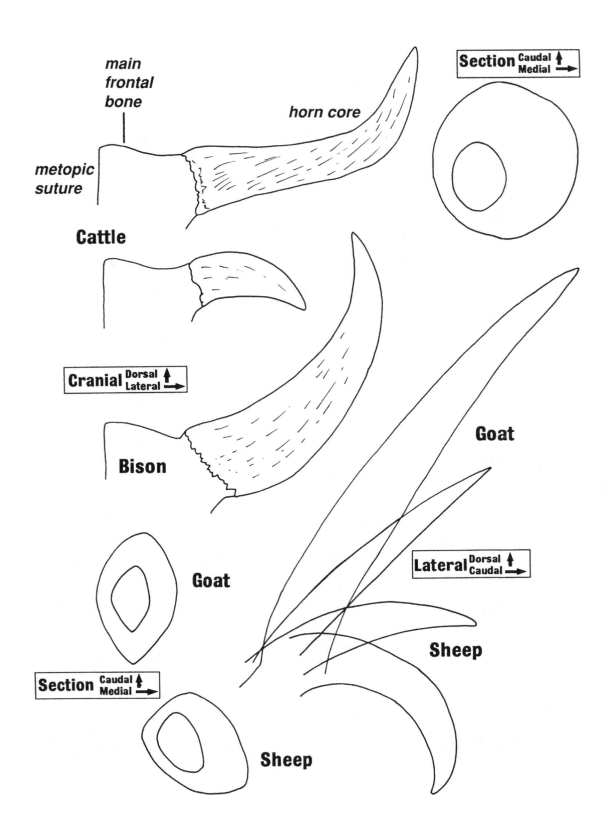

FIGURE 3. HORN CORES.
Upper half: cranial views of the left horn cores in three large bovid specimens - two cattle and one bison - with sections taken just above the attachment to the skull, giving an impression of the maximum and minimum likely to be encountered. Approximately half life-size.
Lower half: lateral outlines of left horn cores in four small bovid specimens - two goats and two sheep - with sections taken just above the attachment to the skull, giving an impression of the maximum and minimum likely to be encountered for both sheep and goat. Approximately half life-size.

quality). Again, specimens prepared from modern carcases are often too large and heavily built by prehistoric standards. By far the largest variation, however, is shown in domestic dogs. A wolf, the wild ancestor of all domestic breeds, is about the size of a German Shepherd dog and has bones not much different in size to those of a large sheep. Some domestic dogs, however, have bones which are almost as small as those of a cat and breeds such as the Great Dane or the Irish Wolfhound have bones as large as one of the small cattle forms. Again, it is size-independent details which must be used for identification.

ANTLERS (FIG. 2)

Antlers are not horns. They are bony outgrowths from the skull which are found only in the cervids. Amongst most deer, only the males grow antlers, but female caribou also bear them (rather smaller). A new set of antlers is grown and shed each year and the size and complexity of the growth increases with each set. There is large variation between individuals and it is difficult to define a "typical" antler. For similar reasons, it is also difficult to use antler development for age estimation.

There is no universal terminology for antlers, but the names below are taken loosely from Taylor Page (1971):

Pedicle – bony protrusion from the frontal bone of the skull, from which the antler arises.

Beam – main axis of the antler onto which all other elements are attached.

Tine – extension from the beam.

Palm – part of the antler which spreads out broadly, usually with points around its edge.

Gutters – irregular grooves which run along the length of the beam and tines.

Pearling – knobbly structure found on the surface of antlers, particularly near the base.

Coronet – knobbly frill which surrounds the attachment to the pedicle.

A shed antler shows a rough, somewhat bulging area to ventral of the coronet. Where the antler has been detached by human hand, the pedicle is cut or broken through and the remnants of this are visible below the coronet.

The following distinctions may be made:
• **Roe deer**. Small, lightly built antlers with a short beam and short tines (usually three per antler), with very prominent pearling and prominent gutters producing a markedly rough surface.
• **Red deer**. Large, massive antlers, with a long heavy beam of rounded section and long pointed tines. The ideal full development is with one tine low over the coronet (the brow tine), a second tine just above it (the bay tine), a third about half way up the beam (the tray tine) and a crown of three points. The surface has heavy pearling and prominent gutters, especially in the ventral half of the beam.
• **Fallow deer**. Intermediate in size and robustness, with a rounded beam section, a prominent and pointed brow tine just above the coronet, one further tine half way up the beam, and a prominent flattened palm in the dorsal part of the antler, with many points on its caudal edge. The smooth surface has little pearling and less pronounced gutters.
• **Caribou**. A long, slender, rather strap-like beam (often oval in section), often sharply angulated halfway along, or sweeping up in a pronounced curve. Most tines have palmated ends. The surface is smooth, with broad and shallow gutters.
• **Moose**. A very short, very stout beam, with no tines arising from it, and a massive, broad and deeply dished palmated area, from the edge of which arise many points. The surface is moderately smooth, with only moderate pearling and gutters.

HORN CORES (FIG. 3)

Bovids have true horns – that is, structures based upon a bony extension of the frontal bone of the skull (the *horn core*) and covered with a sheath of keratin. No part of the horn is shed and the whole structure grows throughout life. All this is in contrast to the skin-covered horns of the Giraffidae and the unique horns of the Antilocapridae, or pronghorn, which are covered in a layer of fused hair which is shed each year. True horns are present in all male bovids, and present in the females of most genera. Females in some breeds of domesticated cattle, sheep and goats are hornless, or *polled*. In such cases the point on the frontal bone where the horn core would have arisen may be marked by a shallow indentation or a low hump, the surface of which may be roughened.

The following distinctions may be made:
• **Cattle**. A massive, round cross section, highly variable in size (especially small in domestic animals). The main distinguishing characteristic is a twist along the length, so that the core sweeps to lateral and to cranial from the skull.
• **Bison**. A similarly massive, round cross section to cattle. The main distinction is that the cores arc directly to dorsal from the frontal bone, with no twist along their length.
• **Goat**. Less robustly constructed, oval in cross section and often with a sharp crest along the cranial border, curving to caudal from the frontal bone, sometimes with a slight lateral twist.
• **Sheep**. Similar in robustness to goat, but usually with a more rounded cross-section and further distinguished by curving spirally to caudal and out to lateral around the ear region of the skull.

CRANIAL BONES (FIG. 2)

Most elements of the skull are readily identifiable but much of their anatomy is too complex to describe in this short guide. There are, however, some basic distinctions. The left and right *frontal bones* form the forehead region in humans and are separated during early infancy by the joint of the *metopic suture*. In all but a very few humans, they fuse together into a single bone very early on in life, but they remain separate throughout adulthood in many mammals, including sheep and goats. Frontal bones are large structures in all bovids and the horn cores are extensions of them. The frontal sinus, which is a small but complex chamber just to dorsal of the nose and eye sockets (or *orbits*) in humans, is an enormously expanded labyrinth in bovids, extending into the horn cores' hollow centres. Conversely, the left and right *parietal bones* which form the top and sides of the human skull, and which frequently remain separate into adulthood, are normally fused together in sheep and goats and are tucked in at the back of the skull, to caudal of the horn cores. On the lateral side of each parietal runs a ridge, the *temporal line*, which marks out the attachment for the temporalis muscle of the jaws. Some of the clearest distinctions between sheep and goats are made in the form of the *coronal suture*, which joins the frontals to the parietal, and of the *lambdoid suture*, which forms the caudal border of the parietal, and the way in which this interacts with the temporal line.

Cervids *vs* bovids
• Male **cervid** skulls (and the female caribou) have pedicles arising from the frontal bone, for the attachment of antlers. **Bovids** (sometimes not including females) have horn cores arising from their frontal bone.
• **Bovid** skulls may or may not have a *lacrymal pit* (an oval depression to cranial of the orbit). **Cervid** skulls always have one.
• **Bovid** skulls usually have only one *lacrymal foramen* at the cranial edge of each orbit (Lawlor 1979). **Cervid** skulls usually have two. There are occasional exceptions to this rule.

Sheep *vs* goat
The following features are distinctive in a large proportion of reference specimens, but some show intermediate characters and a few do not conform at all (Clutton-Brock *et al.*, 1990).
• In cranial view, the coronal suture runs along a relatively straight line in **goats** (Prummel & Frisch, 1986). In **sheep** it usually has a distinct angle about the median sagittal plane.
• The lambdoid suture runs along a relatively straight line in **sheep**. In **goats** it usually curves, bulging to cranial between the left and right temporal lines.

TEETH (FIGS. 4 – 16)

Teeth have two elements, the *crown*, which is coated with hard and glossy enamel and the *roots*. The inner core of both crown and roots is constructed of dentine, softer than enamel but still very resilient when fresh (billiard balls used to be made out of ivory – which is no more or less than elephant dentine). The surface of the root is coated with a bone-like tissue called cement. This cement coating often covers the enamel surfaces of the crown as well. The teeth of horses are particularly thickly invested in a cement jacket. Wear during life may remove the cement from the crown, and it often fractures away from archaeological specimens.

The basic structure of a crown may be seen in the simple cheek teeth of humans, where the occlusal surface is decorated with modest humps called *cusps*. In between the cusps are irregular *fissures*, and the cusps may also be connected by *ridges*. All the other animals described in this key have much higher-crowned teeth than humans do, and the effect of this can be imagined in terms of increased height of cusps and their connecting ridges. In the cat and dog, high cusps and sharp ridges form dagger and blade-like teeth for dispatching and slicing prey. In the pig, the main cusps of the cheek teeth are only a little higher than those of humans, but the complexity of the crown is also increased by the addition of further "accessory" cusps in between. In cattle, sheep, goats and deer, the pig pattern of cusps is greatly increased in height, so that the fissures in between become deep pits (*infundibulum*) extending down from the occlusal surface. The sides of the greatly heightened accessory cusps form pillars and *infoldings* down the vestibular and lingual sides of the crowns. This arrangement is very clear in unworn teeth but, as the crowns of these animals wear rapidly, the tips of the cusps are soon lost. In addition, the whole crown surface is thickly coated with cement, which also fills the infundibulum.

One further complicating factor is that almost all mammals have two dentitions; a *deciduous*, or milk dentition when they are young, and a *permanent* dentition when mature. In general, deciduous teeth are distinguished from permanent by their lower, more chubby and bulging crowns, with a marked waisting in the cervical region. Deciduous teeth roots are relatively slender and, in multirooted teeth, more widely spreading to accommodate the crowns of the permanent teeth which are growing underneath. As the permanent teeth continue to grow, the roots of the deciduous teeth above them are resorbed so that, by the time they are lost, very little of the roots may remain.

The teeth in dentitions are usually summarised as *dental formulae* (Hillson, 1990). For a "generalised mammal" deciduous dentition this would be:

$$\leftarrow\text{mesial} \qquad di\frac{3}{3} \quad dc\frac{1}{1} \quad dp\frac{4}{4} \qquad \text{distal}\rightarrow$$

where *di* is deciduous incisors, *dc* is deciduous canines and *dp* is deciduous premolars. The formula represents the left half of the dentition, viewed as if staring its owner in the face. The median sagittal plane is at the left side, so the deciduous incisors are the most mesial of teeth and the deciduous premolars the most distal. The top row of figures represents the numbers of teeth of the different classes in the upper left half of the dentition, and the lower row of figures similarly represents the numbers of teeth of the different classes in the lower left half of the dentition. The equivalent dental formula for the generalised mammalian permanent dentition would be:

$$i\frac{3}{3} \quad c\frac{1}{1} \quad p\frac{4}{4} \quad m\frac{3}{3}$$

where *i* is permanent incisors, *c* is permanent canines, *p* is permanent premolars and *m* is permanent molars. Teeth in each class are numbered from the most mesial. So the first incisor is the most mesial incisor, the third is the most distal and the second incisor is the one in between. In almost all mammals, both deciduous and permanent dentitions are reduced from this generalised form in some way. Where this has occurred, the teeth in different classes are still numbered as they would be in the full dental formula. So, for example, in humans the permanent premolars are here called the third and the fourth premolars, even though there are only two of them in each quadrant of the dentition.

HUMAN TEETH (FIG. 4)

$$di\frac{2}{2} \quad dc\frac{1}{1} \quad dp\frac{2}{2} \quad and \quad i\frac{2}{2} \quad c\frac{1}{1} \quad p\frac{2}{2} \quad m\frac{3}{3}$$

Human incisors and canines are all approximately spatulate in form, a combination of three tall cusps. The central cusp dominates in the canines, with prominent ridges either side. The three cusps of the incisors are coalesced through much of their height, and their tall sides form the bulk of the tooth (the tiny separate cusp tips rapidly wear away). Human cheek teeth (premolars and molars) are lower crowned than any of the other animals considered in this book. Their broad occlusal surfaces are decorated by low cusps.

PIG TEETH (FIG. 4)

$$di\frac{3}{3} \quad dc\frac{1}{1} \quad dp\frac{3}{3} \quad and \quad i\frac{3}{3} \quad c\frac{1}{1} \quad p\frac{4}{4} \quad m\frac{3}{3}$$

The most distinctive pig teeth are the large, persistently growing permanent canine tusks. By contrast, some other teeth are reduced to tiny points – permanent and deciduous third incisors, and the permanent first premolars in both upper and lower jaws. The permanent and deciduous lower first and second incisors are all long, narrow, spatulate teeth which, when worn, look a little like chisels. The remaining (non-reduced) permanent and deciduous lower premolars all have three cusps arranged in a mesial-distal line to form a high ridge. As for the rest of the cheek teeth, they are all relatively low crowned, having broad occlusal surfaces decorated with cusps (somewhat taller than in human cheek teeth) which wear away rapidly.

BOVID AND CERVID TEETH (FIG. 5)

$$di\frac{0}{3} \quad dc\frac{0}{1} \quad dp\frac{3}{3} \quad and \quad i\frac{0}{3} \quad c\frac{0}{1} \quad p\frac{3}{3} \quad m\frac{3}{3}$$

The lower permanent and deciduous incisors and canines are all spatulate in form, arranged in a fan which bites against a horny pad in the upper jaw. The cheek teeth are widely separated from the canines by a gap (*diastema*), and are all markedly high crowned, with a deep infundibulum and infoldings. They rapidly come into wear and this exposes a complex pattern of enamel ridges in the occlusal surface.

Bison, cattle, sheep and goats are all very similar in the form of their teeth. Cattle and bison can be distinguished from the small bovids by their size. Payne (1985) has suggested some differences between sheep and goat in the lower deciduous third and fourth premolars, and permanent first molar. Cervid cheek teeth are distinguished from bovid by their lower and generally less robust crowns, by their pronounced cervical bulge, by the sharpness and divergence of the accessory pillars on their vestibular side, and by the reduction in the accessory cusp on the lingual side in upper molars and vestibular side in lower molars. The different cervids are difficult to distinguish from one another in their teeth.

HORSE TEETH (FIG. 6)

$$di\frac{3}{3} \quad dc\frac{0}{0} \quad dp\frac{3}{3} \quad and \quad i\frac{3}{3} \quad c\frac{0-1}{0-1} \quad p\frac{3-4}{3} \quad m\frac{3}{3}$$

Both upper and lower incisors are present, and are conical in form, each with a single infundibulum. The lower and upper canines, and upper first premolars are often absent and, even if present, are reduced to small pegs. The upper cheek teeth are all similar to one another, with a tall, parallel-sided crown, double infundibulum and complex infoldings. Lower cheek teeth also are all similar, with again a tall and parallel-sided crown, but with no infundibulum and instead a very complex pattern of infolds. The permanent cheek tooth crowns are persistently growing and their roots never become prominent. Deciduous cheek teeth have more prominent roots and lower crowns, but are otherwise very similar to their permanent successors.

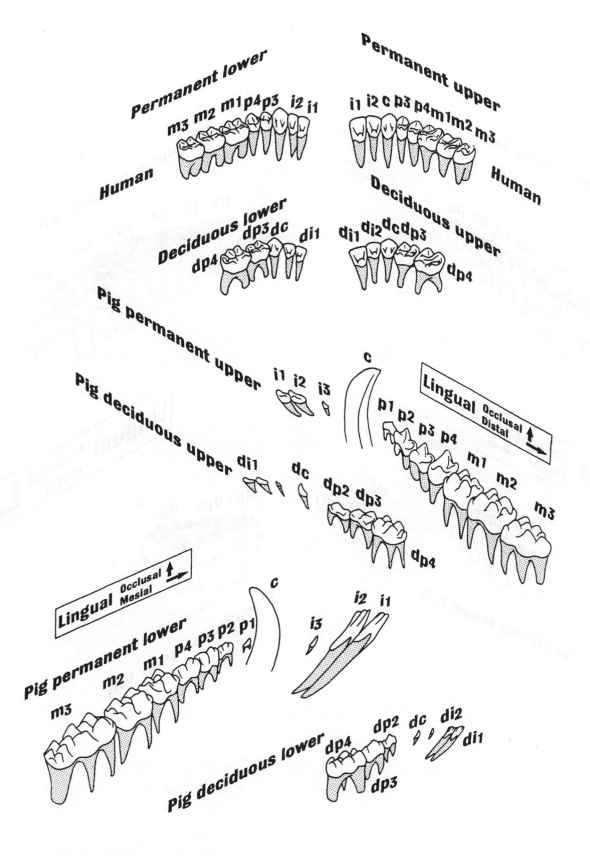

FIGURE 4. DENTITIONS OF HUMAN AND PIG.

Left halves of the upper (on the right hand side of the diagram) and the lower (on the left hand side of the diagram) dentitions. Isometric projection, reproduced at one half life-size.

1st row: human permanent dentition.
2nd row: human deciduous dentition.
3rd row: pig permanent upper dentition.
4th row: pig deciduous upper dentition.
5th row: pig permanent lower dentition.
6th row: pig deciduous lower dentition.

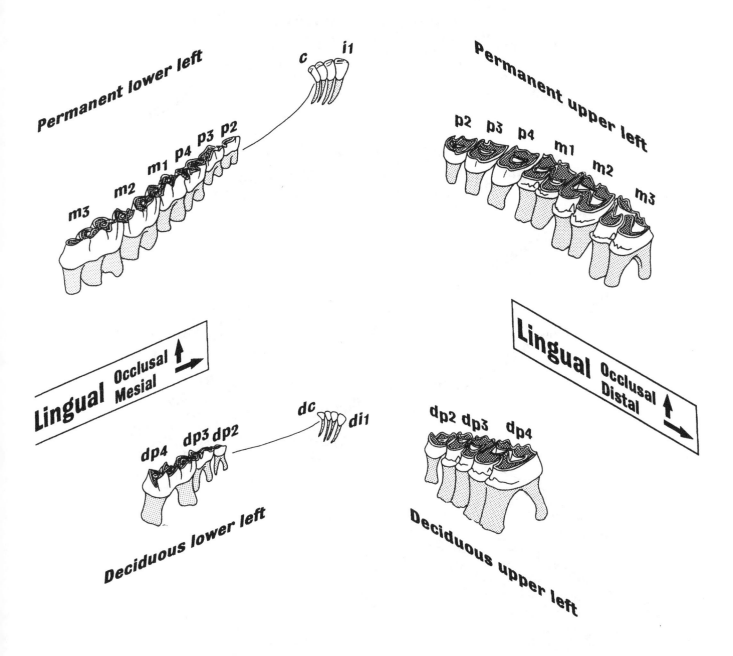

FIGURE 5. DENTITIONS OF RED DEER.
The red deer has been chosen to represent all the bovids and cervids, because its teeth are roughly intermediate in size. Left halves of the upper (on the right hand side of the diagram) and the lower (on the left hand side of the diagram) dentitions. Isometric projection, reproduced at one half life-size.
1st row: permanent dentition.
2nd row: deciduous dentition.

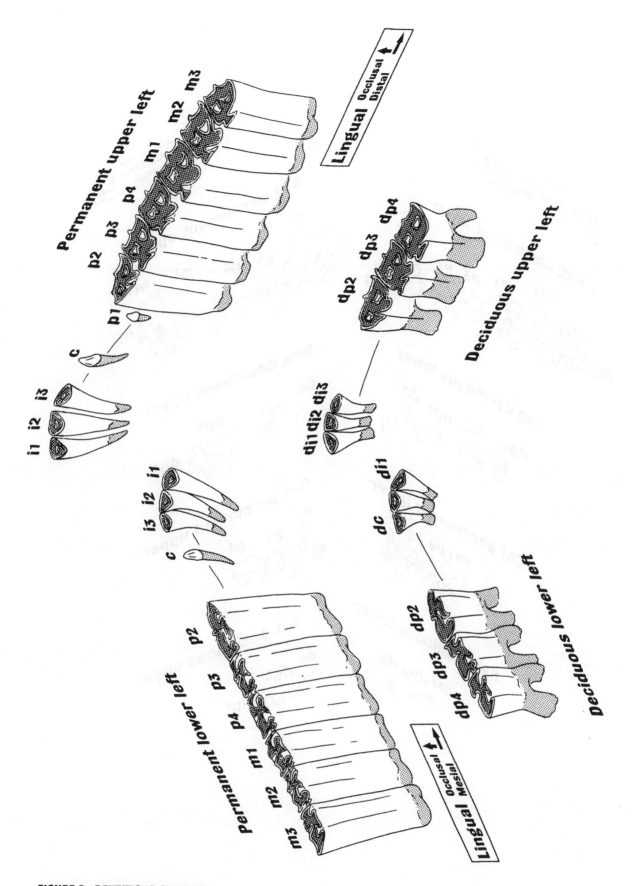

FIGURE 6. DENTITIONS OF HORSE.

Left halves of the upper (on the right hand side of the diagram) and the lower (on the left hand side of the diagram) dentitions. Isometric projection, reproduced at one half life-size.

1st row: permanent dentition.

2nd row: deciduous dentition.

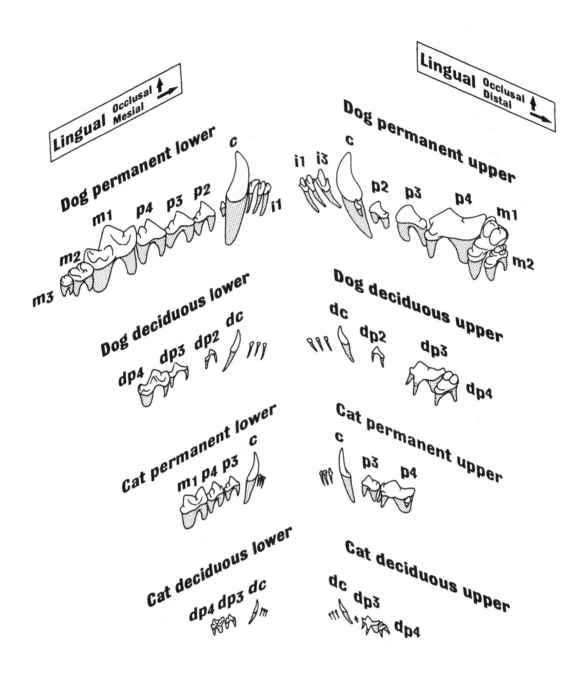

FIGURE 7. DENTITIONS OF DOG AND CAT.

Left halves of both the upper (on the right hand side of the diagram) and the lower (on the left hand side of the diagram) dentitions. Isometric projection, reproduced at one half life-size.

1st row: dog permanent dentition.

2nd row: dog deciduous dentition.

3rd row: cat permanent dentition.

4th row: cat deciduous dentition.

Dog $\quad di\frac{3}{3} \quad dc\frac{1}{1} \quad dp\frac{3}{3} \quad$ and $\quad i\frac{3}{3} \quad c\frac{1}{1} \quad p\frac{4}{4} \quad m\frac{2}{3}$

Cat $\quad di\frac{3}{3} \quad dc\frac{1}{1} \quad dp\frac{3}{2} \quad$ and $\quad i\frac{3}{3} \quad c\frac{1}{1} \quad p\frac{2}{2} \quad m\frac{1}{1}$

The dentitions of these carnivores are greatly modified from the generalised mammal dental formula. Permanent first premolars in the dog are reduced to small points and these teeth, together with the second premolar, are not even present in the cat. All the incisors in both animals are spatulate in form, with three cusps arranged in a somewhat fleur-de-lys form, and packed together into a comb-like structure at the front of the jaw. Upper and lower canines are widely separated from the other teeth, and have tall, robust, pointed crowns and long, stout roots. Most of the remaining premolars in cat and dog are formed into distal-facing hooks, but the upper permanent fourth and deciduous third premolars have a special function. Together with the lower permanent first molar and deciduous fourth premolar, they are modified into blade-like *carnassial* teeth which cut against one another like scissors. This blade-like form is more pronounced in cats than in dogs. The remaining molars in the dog are low-cusped crushing teeth, but in the cat they are much reduced or missing.

THE TEETH DIAGRAMS

Most diagrams of teeth in this book have been drawn using an isometric projection. They are an attempt to represent, in outline, the intricate three-dimensional structure that you might see holding a tooth specimen in your hand. All are drawn as though the tooth is being held by its roots, with the occlusal surface of the crown uppermost. One of the two main viewpoints used in the diagrams is from above the mesial/vestibular/occlusal corner. In this case, the isometric projection makes it possible to look down upon the occlusal surface of the crown, and along the mesial and vestibular sides. The other main viewpoint is from above the distal/lingual/occlusal corner, and the viewer looks down similarly upon the occlusal surface and along the distal and lingual sides. In some drawings, both views are given for one tooth and, where this has been done, a box connects the two. Where it is practicable, teeth are shown both in the unworn state, and in varying states of wear. The process of wear exposes the underlying dentine, and this is indicated by shading with a grey tone. Roots are shaded with a lighter tone.

IDENTIFYING TEETH

If reduced and peg-like teeth are ignored, the teeth of animals covered in this book can be split in terms of their general form into nine groups:

GROUP 1 – SPATULAE

These teeth have crowns initially with three cusps arranged in a mesial-distal line, forming a spatulate or chisel-like structure with its convex surface to vestibular and more concave surface to lingual. Wear soon removes the cusps to accentuate the spatulate form of the crown. All are single rooted.

General left-right distinctions

The following features allow the tooth to be orientated:
• The crown is asymmetrical and tilts to distal, the tilt increasing along the tooth row so that the most mesial tooth is most symmetrical and the most distal tooth is most asymmetrical.
• The cervical margin of the crown curves up to occlusal on the mesial and distal sides, the mesial curve being slightly more pronounced.
• The root curves to distal, least on the most mesial tooth in the row, increasing steadily to the most distal tooth.

1a. Human permanent and deciduous upper and lower incisors (Fig. 8)

The spatulate form is particularly prominent in these teeth.

• **Permanent upper first and second incisors.** Crown: broad mesial-distal relative to its height cervical-occlusal, and with a marked asymmetry (especially in the second incisor). Root: stout, with a rounded triangular section.

• **Deciduous upper first and second incisors.** Crown: similar to the permanent upper incisors, but smaller and squatter, and markedly waisted in the cervical region. Root: thin and slender relative to the permanent teeth.

• **Permanent lower first and second incisors.** Crown: similar to the upper permanent incisors, but relatively narrow mesial-distal and with less marked asymmetry. Root: more slender than in the upper incisors, and oval in section.

• **Deciduous lower first and second incisors.** Crown: similar to the permanent lower incisors, but smaller and squatter. Root: relatively thin and slender.

1b. Human permanent and deciduous upper and lower canines (Fig. 8)

• **Permanent upper canine.** Crown: tall cervical-occlusal, stoutly bulging mesial-distal, with the central cusp the dominant element, forming a broad buttress down the centre of the lingual side. Root: stout, long, rounded triangular section.

• **Deciduous upper canine.** Crown: similar to the permanent upper canine, but squatter and more markedly bulging out mesial-distal from a narrow cervical region. Root: thin and slender relative to the permanent tooth roots.

• **Permanent lower canine.** Crown: as tall cervical-occlusal as the upper permanent canine, but narrower mesial-distal, with the lingual buttress less prominent and the spatulate form more pronounced. Occasionally, worn crowns may be confused with the permanent upper first incisor. Root: stout, long, oval in section.

• **Deciduous lower canine.** Crown: similar to the lower permanent canine, but squatter and more bulging. Root: relatively thin and slender.

1c. Bovid and cervid permanent and deciduous lower incisors and canines (Fig. 8)

All these teeth are similar in form.

• **Permanent incisors and canine.** Crown: somewhat similar to the human upper incisors, but with a more flaring crown, narrower cervical region, and a much more pronounced asymmetry (greatly increasing from mesial to distal along the tooth row). Root: round section, strongly curved to distal (curve increasing from first incisor to canine).

• **Deciduous incisors and canine.** Crown: similar to the permanent teeth, but shorter cervical-occlusal and narrower cervically, giving a squatter, more flaring vestibular outline. Root: similar to the permanent tooth, but relatively thin and slender.

1d. Dog and cat permanent and deciduous upper and lower incisors (Fig. 9)

• **Permanent incisors.** Crown: spatulate in form, but with the central cusp more prominent and clearly separated from the mesial and distal cusps, little worn crowns sometimes taking a *fleur-de-lys* form, and sometimes with the mesial and distal cusps being placed more to lingual and enclosing a small pit to lingual of the central cusp. The mesial cusp is often coalesced with the central cusp in the lower incisors. Root: single, with a rounded to oval section.

• **Deciduous incisors.** Crown: similar in form to the permanent teeth, but reduced to a large extent. Root: very narrow and slender.

• **Cat incisors** are very much smaller than those of even the smallest dog, and the deciduous teeth are reduced to mere pinheads.

1e. Pig permanent and deciduous lower incisors (Fig. 8)

• **Permanent first and second incisors.** Crown: relatively deep vestibular-lingual, with the central cusp prominent as a buttress for the whole cervical-occlusal height, giving the crown a chisel-like form. The cervical margin on the mesial and distal sides curves markedly to occlusal. Root: single, long, stout, rounded in section.

• **Deciduous first and second incisors.** Crown: very like the permanent tooth, but smaller and relatively shorter cervical-occlusal. Root: narrower than the permanent roots.

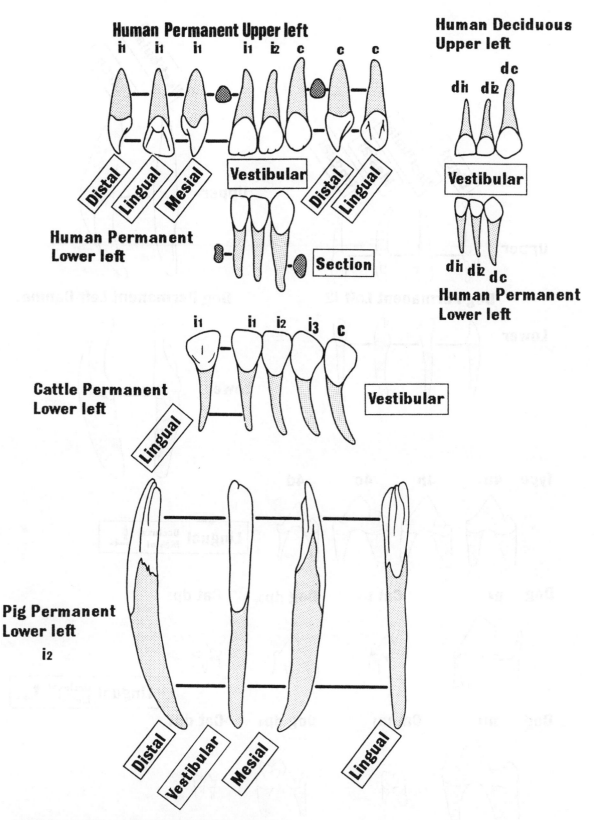

FIGURE 8. SPATULATE HUMAN, CATTLE AND PIG TEETH.

1st row: human upper left permanent and deciduous incisors and canines (Groups 1a and 1b in the text), in the following order - first incisor seen in distal, lingual and mesial views, and in root section; first and second incisor, and canine, seen in vestibular view; canine seen in root section, distal and lingual views; deciduous incisors and canine seen in vestibular view.

2nd row: human lower left permanent and deciduous incisors and canines (Groups 1a and 1b in the text), in the following order - first incisor in root section; first and second incisors, and canine, in vestibular view, canine in root section; deciduous incisors and canine in vestibular view.

3rd row: large bovid (cattle) permanent incisors and canines (Group 1c in the text) - first incisor in lingual view; first, second and third incisors, and canine, in vestibular view.

4th row: pig permanent left lower second incisor (Group 1e in the text)- in distal, vestibular, mesial and lingual views.

All reproduced at approximately life-size.

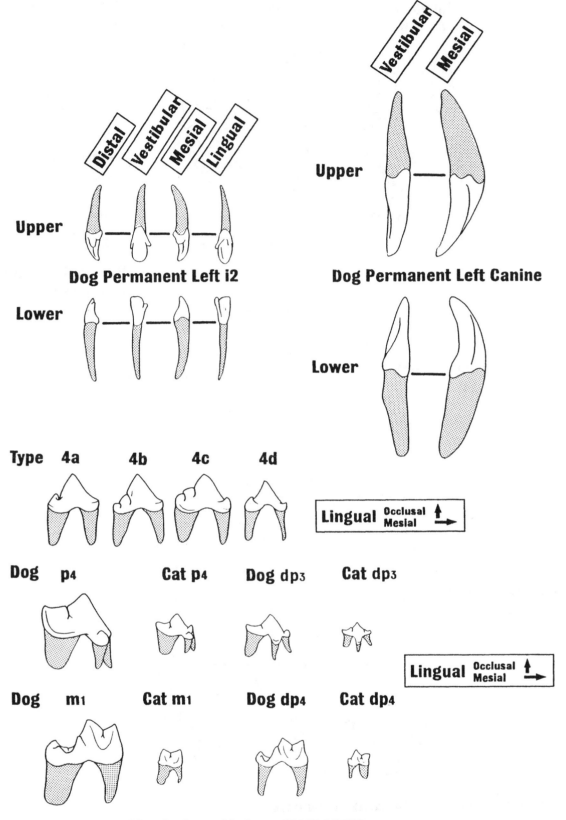

Upper

Dog Permanent Left i2

Upper

Dog Permanent Left Canine

Lower

Lower

Type 4a 4b 4c 4d

Lingual | Occlusal ↑ Mesial →

Dog p4   **Cat p4**   **Dog dp3**   **Cat dp3**

Lingual | Occlusal ↑ Mesial →

Dog m1   **Cat m1**   **Dog dp4**   **Cat dp4**

FIGURE 9. DOG AND CAT SPATULA, POINT, HOOK AND BLADE-LIKE TEETH.

1st row: dog upper left permanent second incisor and canine (Groups 1d and 2 in the text) - first incisor in distal, vestibular, mesial and lingual views; canine in vestibular and mesial views.

2nd row: dog lower left permanent second incisor and canine (Groups 1d and 2 in the text) - first incisor in distal, vestibular, mesial and lingual views; canine in vestibular and mesial views.

3rd row: dog and cat hook-like premolar teeth - groups 4a, 4b, 4c and 4d in the text.

4th row: dog and cat upper carnassials (Group 5 in the text) - dog and cat permanent upper fourth premolar, dog and cat deciduous upper third premolar.

5th row: dog and cat lower carnassials (Group 5 in the text) - dog and cat permanent lower first molar, dog and cat deciduous lower fourth premolar.

All reproduced at approximately life-size.

16

GROUP 2 – POINTS

High, pointed crowns, usually with one distal ridge, and bulging in the cervical region. A single long and stout root. The following features allow the tooth to be orientated:

Distinctions between left and right

The crowns are grooved along their length on the mesial/lingual side. The cervical margin curves to occlusal on the mesial side, where the tooth lies next to the third incisor (the main arc of the tooth is in a cranial-caudal plane, rather than vestibular-lingual).

Distinctions between upper and lower

The lower teeth are somewhat more slender than the upper and their crowns are twisted more outwards to vestibular.

Dog and cat permanent and deciduous upper and lower canines

• **Dog permanent canines.** Crown: tall and pointed, but stout and bulging in the cervical region. Root: long but stout.
• **Cat permanent canines.** Crown: smaller than dog, but also more slender relative to their cervical-occlusal height. Root: smaller and more slender than dog.
• **Dog or cat deciduous canines.** Crown: similar in form to the permanent teeth, but more slender and much lower. Root: shorter and less robust than in the permanent teeth.

GROUP 3 – PIG TUSKS (FIG. 4)

The crown of the permanent pig canine forms a large, curving tusk of triangular section, usually with no root, as the tooth grows continuously. The size varies a great deal, between males and females, and between different populations. This variability often makes it difficult to distinguish between upper and lower canines.

GROUP 4 – HOOKS (FIG. 9)

Amongst the carnivores, many of the pre–molars bear a crown with a relatively high blade and a pointed main cusp which hooks strongly to distal. Much lower accessory cusps lie to mesial and to distal, and these are the basis for identification of individual teeth. All have two roots.

4a. Dog permanent upper second and third premolars

Crown: large and stout, with a single distal accessory cusp only. Roots: robust and moderately spread.

4b. Dog permanent lower second to fourth premolars

Crown: large and stout, with two distal accessory cusps only. Roots: robust and moderately spread.

4c. Cat permanent upper third premolar or lower third and fourth premolars

Crown: small and lightly built, with one small mesial accessory cusp and two distal accessory cusps. Roots: smaller than dog, but still robust relative to crown size.

4d. Dog deciduous upper and lower second premolars

Crown: small, slender vestibular-lingual, with one or two distal accessory cusps, cervical margin arching high to occlusal between the roots. Roots: narrower in section and more widely spread than the permanent roots.

GROUP 5 – BLADES (FIG. 9)

In the carnassial teeth of carnivores, the crown forms a prominent blade which continues at a high level along the crests of two tall cusps. There are two or three stout roots.

5a. Dog and cat permanent upper fourth premolars

• **Dog.** Crown: the stout main mesial cusp rises to a point and then becomes blade-like in its distal portion. The blade is continued along the slightly lower distal cusp. A low rounded cusp lies on the mesial/lingual corner of the crown. Roots: three, one under each cusp, the most distal being particularly broad mesial-distal.
• **Cat.** Crown: not unlike dog, but with a more slender main cusp and a prominent mesial accessory cusp. Roots: three, as in dog.

5b. Cat or dog deciduous upper third premolar
Crown: like the permanent fourth premolar but more slender and with a strongly arched cervical margin between the roots. A very low cusp forms a tongue-like extension just to lingual of the main cusp. In **dog** there is a low mesial accessory cusp with two lobes. In **cat**, this accessory cusp is larger and more strongly divided into two lobes. Roots: three slender roots, widely spread, with the lingual root in a markedly different position to the permanent fourth premolar.

5c. Dog lower permanent first molar and deciduous fourth premolar
• **Permanent first molar.** Crown: two main cusps, the more mesial being slightly less prominent, together forming a blade-like mesial two-thirds of the crown. A lower accessory cusp is sited to lingual of the distal main cusp. The distal one-third of the crown is arranged like a low 'heel' and bears three small cusps. Roots: two stout roots.
• **Deciduous fourth premolar.** Crown: like the permanent first molar in general form, but smaller and more slender vestibular-lingual. The cervical margin is more markedly arched to occlusal between roots and the accessory cusp to lingual of the distal main cusp is more prominent. Roots: two slender, more widely spaced roots.

5d. Cat lower permanent first molar and deciduous fourth premolar
• **Permanent first molar.** Crown: two main cusps of roughly equal prominence, together forming a markedly blade-like crown. There is only a tiny distal accessory cusp, representing the last vestige of the 'heel' seen in the equivalent dog carnassial (Group 5c). Roots: two stout roots, the distal being stouter than the mesial.
• **Deciduous fourth premolar.** Crown: like the permanent first molar, but smaller and more slender vestibular-lingual, with a more marked occlusal arching of the cervical margin between the roots. The main cusps are also less equal in size and the distal accessory cusp is more pronounced. Roots: two slender, more widely spaced roots, the mesial being stouter than the distal.

GROUP 6 – PIG, BOVID AND CERVID TEETH WITH HIGH CROWNS AND SIMPLE INFOLDINGS (FIG. 10)

These relatively tall crowns are elongated mesial-distal, and have one central main cusp connected by a high ridge to slightly lower cusps to mesial and distal. They are decorated by infoldings of varying depth – one infolding from the vestibular side and two (usually deeper) infoldings from lingual. There are two roots, one mesial and one distal. Wear starts early in life and progressive wear of the occlusal surface cuts ever deeper sections through the crown and exposes a changing outline of upstanding enamel ridges. The similar general layout of the equivalent teeth in pigs and bovids or cervids is quite plain in unworn crowns, but becomes less obvious in worn teeth.

6a. Pig lower permanent second to fourth premolars or deciduous second and third premolars
• **Permanent premolars.** The unworn crown is quite high, narrow and blade-like. This impression is emphasised by the very shallow infoldings. Wear initially exposes just the outlines of the three cusps and then, because these merge together at their bases, an oval outline is eventually produced. The crowns form a series of increasing size and increasing depth of infoldings from mesial to distal along the row.
• **Deciduous second and third premolars.** The crowns are little different to the permanent premolars, but are more waisted in the cervical region and have more slender, more spreading roots.

6b. Bovid or cervid lower permanent second to fourth premolars or deciduous second and third premolars
• **Permanent premolars.** Unworn, the crown has a high ridge, somewhat higher than that of pig, with more pronounced lingual infoldings. These may be so deep that they enclose an infundibulum near the cervical region of the crown, particularly in the permanent fourth premolar. Wear exposes first of all the outline of the high ridge, then the "buttress" that divides the two main lingual infoldings and then, if the tooth includes an infundibulum, this eventually becomes isolated as an island in the worn surface. The complexity of these worn surfaces increases markedly from mesial to distal in the tooth row.

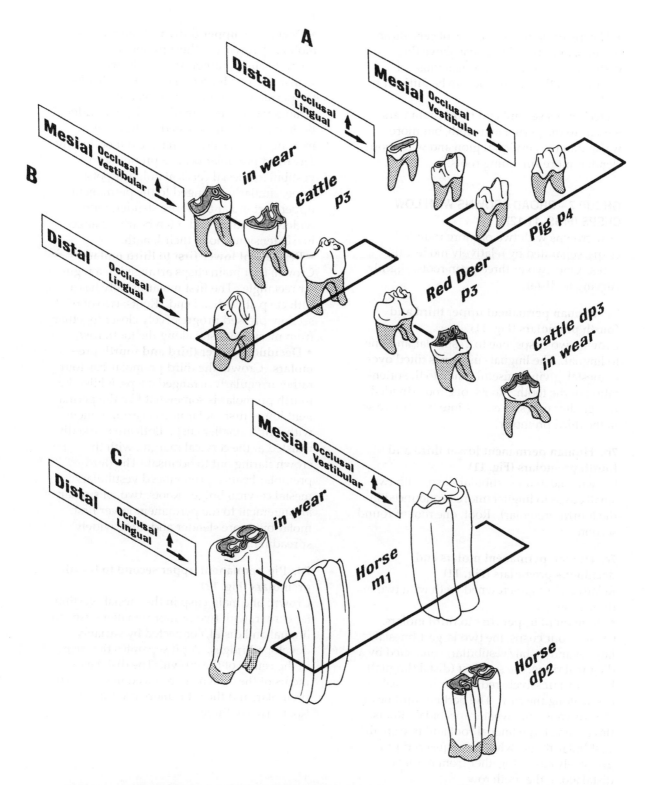

FIGURE 10. HIGH CROWNED TEETH WITH INFOLDINGS.

A. Pig lower left permanent third premolar (Group 6a in the text). Left hand diagonal row shows distal/lingual views of the third premolar (unworn and with two stages of wear). Right hand drawing shows a mesial/vestibular view of the same tooth (unworn).

B. Bovid (cattle) and cervid (red deer) lower left permanent third premolar and deciduous third premolar (Group 6b in the text). Left hand drawing shows a distal/lingual view of a cattle permanent third premolar (unworn). Right hand diagonal row shows mesial/vestibular views of cattle permanent third premolar (unworn and in two stages of wear), red deer permanent third premolar (unworn and in wear), and cattle deciduous third premolar (in wear).

C. Horse lower left permanent first molar and deciduous second premolar (Group 8 in the text). Left hand diagonal row shows distal/lingual views of horse permanent first molar (unworn and with slight wear) and deciduous second premolar (in wear). Right hand drawing shows a mesial/vestibular view of the same horse first molar.

Isometric projection, reproduced at about life-size.

• The lower premolar crowns of **cervids** are somewhat more bulging just above the vestibular cervical margin than those of **bovids**, but the distinction can be difficult to make.

• **Deciduous second and third premolars**. Similar to the permanent teeth, but more waisted in the cervical region and with more slender, more spreading roots.

GROUP 7 – BROAD CROWNS WITH LOW CUSPS (FIGS. 11-12)

Low crowns with two or more rounded cusps, separated by relatively modest fissures. One, two or three main roots, slightly curving to distal.

7a. Human permanent upper third and fourth premolars (Fig. 11)

Crown: two cusps, one to vestibular and one to lingual. The lingual cusp tip is tilted over to mesial, giving a useful guide to the orientation of the tooth. Roots: one root, divided to a greater or lesser extent into two (more so in the third premolar).

7b. Human permanent lower third and fourth premolars (Fig. 11)

Crown: one main vestibular cusp, with two small cusps to lingual (more pronounced in the fourth premolar). Root: one root of round section.

7c. Human permanent molars and deciduous premolars (Fig. 11)

Relatively square-crowned teeth with two or three roots.

• **Permanent upper first to third molars**. Crown: four cusps, the two largest (mesial/lingual and distal/vestibular) connected by a diagonal ridge, the smallest (distal/lingual) being progressively reduced from mesial to distal along the row. The occlusal outline of the crown is somewhat trapezoidal. Roots: three – one large lingual root and two smaller vestibular roots, which are squeezed progressively closer together from mesial to distal along the tooth row.

• **Deciduous upper third and fourth premolars**. Crown: the third premolar has four separate cusps, rather irregularly arranged, whilst the fourth premolar is somewhat like the permanent upper first molar in cusp arrangement, with a marked diagonal ridge. Both are markedly waisted at the cervical margin, with the crown flaring out occlusally. The third premolar bears a pronounced vestibular/mesial cervical bulge. Roots: three, similarly arranged to the permanent upper first molar but more slender, more widely spreading and often bearing flange-like extensions along their length.

• **Permanent lower first to third molars**. Crown: four main cusps arranged in a regular rectangle. The first molar usually has a fifth cusp at its distal end. Roots: two of oval section, pressed progressively closer together from mesial to distal along the tooth row.

• **Deciduous lower third and fourth premolars**. Crown: the third premolar has four rather irregularly arranged cusps, whilst the fourth premolar is somewhat like the permanent lower first molar in cusp arrangement, with a fifth smaller cusp. Both are markedly waisted at the cervical margin, with the crown flaring out to occlusal. The third premolar bears a pronounced vestibular/mesial cervical bulge. Roots: two, of similar arrangement to the permanent lower first molar but more slender and more widely spreading.

7d. Pig permanent upper second to fourth premolars (Fig. 12)

Crown: one main cusp in the mesial/vestibular corner, with two or more smaller cusps to lingual and distal, connected by variably developed ridges. A pit separates the cusps in the centre of the crown. The distal elements of the crown are reduced in the fourth premolar, and the pit is more prominent. Roots: two to three.

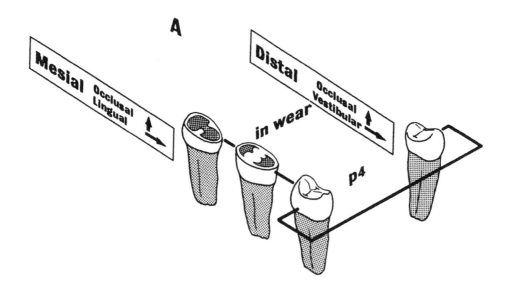

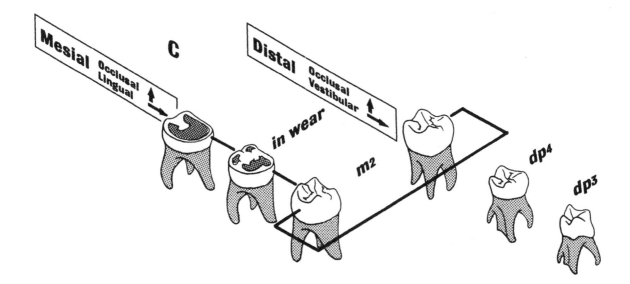

FIGURE 11. TEETH WITH LOW CUSPS (PART 1).

A. Human permanent upper premolar teeth (Group 7a in the text). Left hand diagonal row shows mesial/lingual views of the upper left fourth premolar (unworn and with two stages of wear). Right hand row shows a distal/vestibular view of the same tooth (unworn).

B. Human permanent lower premolar teeth (Group 7b in the text). Left hand diagonal row shows a distal/lingual view of the lower left fourth premolar (unworn). Right hand diagonal row shows mesial/vestibular views of the same tooth (unworn and with two stages of wear).

C. Human upper permanent molar and deciduous premolar teeth (Group 7c in the text). Left hand diagonal row shows mesial/lingual views of the upper left second molar (unworn and with two stages of wear). Right hand row shows a distal/vestibular view of the same left second molar, with upper deciduous fourth and third premolars (all unworn).

D. Human lower permanent molar and deciduous premolar teeth (Group 7c in the text). Left hand diagonal row shows a distal/lingual view of the lower permanent left second molar (unworn). Right hand row shows mesial/vestibular views of the same permanent molar, the lower permanent left first molar and the lower deciduous fourth premolar (all unworn).

Isometric projections, reproduced at about life-size.

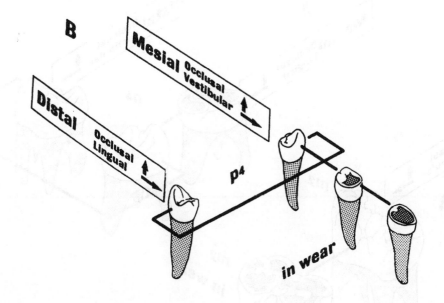

B

Mesial Occlusal Vestibular ↑→

Distal Occlusal Lingual ↑→

p4

in wear

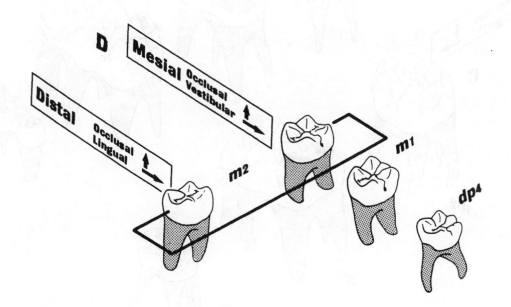

D

Mesial Occlusal Vestibular ↑→

Distal Occlusal Lingual ↑→

m2

m1

dp4

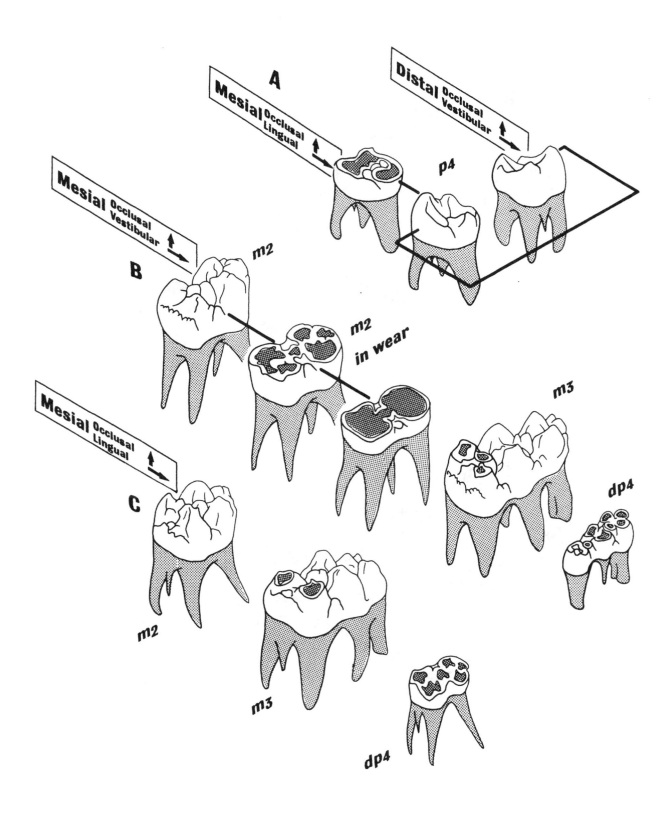

FIGURE 12. TEETH WITH LOW CUSPS (PART 2).

A. Pig upper left permanent fourth premolar (Group 7d in the text). Left hand diagonal row shows mesial/lingual views of the fourth premolar (worn and unworn). Right-hand drawing shows a distal/vestibular view of the same tooth (unworn).

B. Pig lower permanent molars, and deciduous premolar (Groups 7e and 7f in the text). The diagonal row shows mesial/vestibular views of the lower left second molar (unworn and in two stages of wear), the lower left third molar (with some wear) and the lower left deciduous fourth premolar (worn).

C. Pig upper left permanent molars, and upper left deciduous fourth premolar (Groups 7e and 7f in the text). The diagonal row shows mesial/lingual views of the left upper second molar (unworn and in two states of wear), the upper third molar (with some wear) and the upper deciduous fourth premolar (worn).

Isometric projection, reproduced at about life-size.

FIGURE 13. TEETH WITH LOW CUSPS (PART 3).

Dog upper and lower permanent left molars (Group 7g in the text). Upper drawings show mesial/lingual and distal/vestibular views of the upper first molar (unworn). Lower drawings show distal/lingual and mesial/vestibular views of the lower second molar (unworn). Isometric projection, reproduced at about life-size

7e. Pig upper and lower permanent first and second molars or upper deciduous fourth premolar (Fig. 12)

Crowns with four relatively tall main cusps and lower accessory cusps, and four main roots.

• **Permanent upper first and second molars.** Crown: relatively square in occlusal outline, with four main cusps (tall compared with human molars) arranged in a square, with a lower accessory cusp in the centre and further accessory cusps along the distal edge. The surface of the cusps is rougher, and their outline more irregular, than in human molar crowns (Group 7c). Wear progresses rapidly and the outlines of the cusps are soon lost. Roots: four, with short accessory rootlets between the mesial and distal root pairs.

• **Deciduous upper fourth premolar.** Similar to the permanent upper first molar in size and form, but more waisted in the cervical region and with four more slender and more spreading roots.

• **Permanent lower first and second molars.** Crown: a similar arrangement to the upper molars, but rather longer mesial-distal and narrower vestibular-lingual. Roots: four.

7f. Pig permanent third molar or deciduous lower fourth premolar (Fig. 12)

Crowns with six main cusps, markedly elongated mesial-distal.

• **Permanent upper third molar.** Crown: similar to the upper second molar (Group 7e), but with an additional distal element comprising two slightly smaller main cusps. Roots: five main roots, with one or more accessory rootlets.

• **Permanent lower third molar.** Crown: the main part of the crown is similar in form to the lower second molar, but a distal element is added, with two further main cusps (a little smaller than the more mesial four) and further accessory cusps to produce the longest (mesial-distal) crown in the dentition. Roots: five.

• **Deciduous lower fourth premolar.** Crown: six main cusps arranged, in pairs, in a long rectangle (increasing slightly in size from mesial to distal), separated by two central cusps and with a distal accessory

cusp. The cervical margin is deeply indented between the pairs of main cusps. Roots: two large and spreading roots, to mesial and distal, with one slender vestibular root.

7g. Dog permanent upper molars, or dog and cat deciduous upper fourth premolar (Fig. 13)

• **Dog first and second upper molars.** Crown: two broadly bulging vestibular cusps, with a much lower lingual portion to the crown made up of sinuous ridges. The first molar is much larger than the second. Roots: three (one lingual and two vestibular).

• **Dog and cat deciduous upper fourth premolar.** Crown: similar to the dog permanent first molar, but much smaller and somewhat simplified (especially in the cat). Roots: three (one lingual and two vestibular).

7h. Dog permanent lower second molar (Fig. 13)

Crown: low, with three main cusps, looking somewhat like the low distal portion of the dog permanent lower first molar (Group 5c). Roots: two, often pressed together.

GROUP 8 – TALL CROWNS WITH COMPLEX INFOLDINGS

The lower cheek teeth of horses have high crowns and a complex pattern of infoldings from the vestibular and lingual sides, with two short roots.

Horse lower permanent second to fourth premolars and first to third molars, or deciduous second to fourth premolars (Fig. 10)
• **Permanent premolars and molars.** Crown: tall, with a rectangular occlusal outline, roughly parallel-sided, with three small infoldings to vestibular, two deep and two shallow infoldings to lingual. A complex occlusal pattern of enamel ridges is exposed early on in wear and remains similar throughout the functional life of the tooth. All teeth in the close-packed row are very similar, except for the second premolar and third molar which form the ends of the row and are triangular in occlusal outline (Fig. 6). Roots: the crown continues to grow for some time, so that roots appear only at the last stages of tooth formation. At this point, there are two short, stout roots per tooth.
• **Deciduous premolars.** Crown: very similar in the occlusal outline and pattern of enamel ridges to the permanent teeth, but with much lower crowns, more waisted at the cervical margin. Roots: much more prominent than in the permanent teeth, and more widely spreading. Two per tooth.

GROUP 9 – TALL CROWNS WITH AN INFUNDIBULUM

High crowns with single, double or triple infundibulum, combined with infoldings to make a complex occlusal outline. One, two or three main roots.

9a. Horse incisors (Fig. 14)
• **Permanent upper and lower first to third incisors.** Crown: gradually expanding from cervical to occlusal into a trumpet-like form, 'D'-shaped in section, with a single infundibulum extending down inside the crown for much of its height. On wear, the infundibulum is exposed as an 'island' in the occlusal surface. Root: single, appearing late in the development of the tooth, not clearly distinguishable from the crown at the surface.
• **Deciduous upper and lower first to third incisors.** Similar to the permanent teeth, but with a more flaring crown (particularly to mesial and distal) and more discernible roots.

9b. Bovid or cervid permanent upper premolars or deciduous upper second premolar (Fig. 14)
• **Permanent upper second to fourth premolars.** Crown: 'D'-shaped occlusal outline, with a single infundibulum. Two broad infolds from the vestibular side separate three ridges down the vestibular surface. In the unworn state, the crown has two main cusps, connected by ridges which surround the single infundibulum. Wear proceeds rapidly and the infundibulum is soon isolated as an island in the 'D'-shaped occlusal surface. Roots: three (one lingual and two vestibular).
• The crowns of **cervids** show a prominent cervical bulge on the lingual side, so that the crown tapers to occlusal. They also have sharper, more flaring ridges on the vestibular surface than do bovids.
• **Deciduous upper second premolar.** Similar in form to the permanent teeth in both bovids and cervids, but with a lower crown and only two roots (Figs. 5, 17).

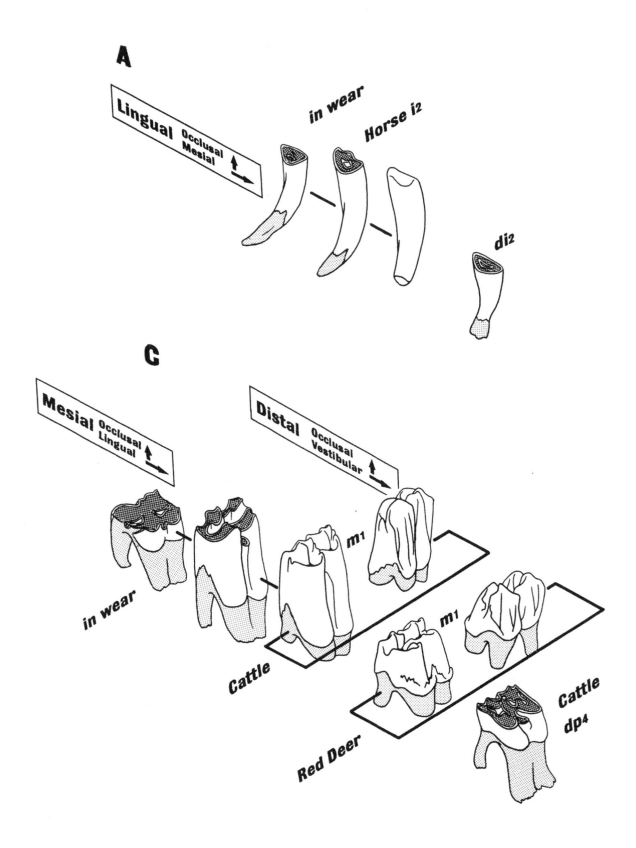

FIGURE 14. TEETH WITH AN INFUNDIBULUM (PART 1).

A. Horse permanent and deciduous lower left incisors (Group 9a in the text), showing lingual/mesial views of a horse permanent second incisor (worn and in two stages of wear), and deciduous second incisor (in wear).

B. Bovid (cattle) and cervid (red deer) upper left permanent third premolars (Group 9b in the text). Left hand diagonal row shows mesial/lingual views of a cattle third premolar (unworn and in two stages of wear) and red deer third premolar (unworn). Right hand row shows distal/vestibular views of the same cattle third premolar and red deer third premolar (both unworn).

C. Bovid (cattle) and cervid (red deer) upper left permanent second molars and deciduous fourth premolars (Group 9c in the text). Left hand diagonal row shows mesial/lingual views of a cattle second molar (unworn and in two stages of wear), a red deer second molar (unworn), and cattle deciduous fourth premolar (worn). Right hand row shows distal/vestibular views of the same cattle and red deer second molar (unworn). Isometric projections, reproduced at approximately life-size.

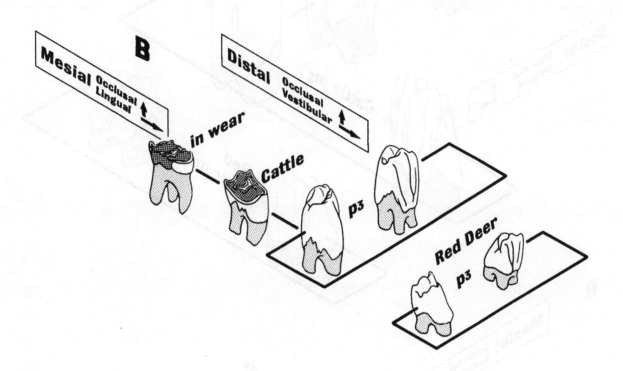

B

Mesial Occlusal
Lingual ↑
→

in wear

Cattle

Distal Occlusal
Vestibular ↑
→

p3

Red Deer

p3

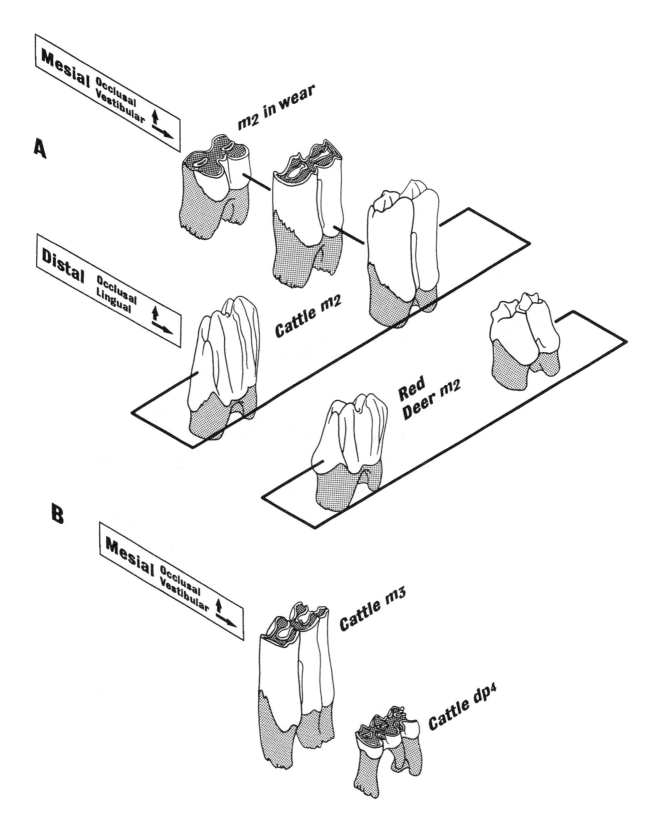

FIGURE 15. TEETH WITH AN INFUNDIBULUM (PART 2)

A. Bovid (cattle) and cervid (red deer) lower left permanent second molars (Group 9c in the text). Left hand diagonal row shows distal/lingual views of cattle second molar and a red deer second molar (both unworn). Right hand row shows mesial/vestibular views of the same cattle second molar (unworn and in two stages of wear) and red deer second molar (unworn).

B. Bovid (cattle) lower left permanent third molar and deciduous fourth premolar (Group 9d in the text). Left hand drawing shows a mesial/vestibular view of the permanent third molar. Right hand drawing shows a mesial/vestibular view of the deciduous fourth premolar.

Isometric projections, reproduced at about life-size.

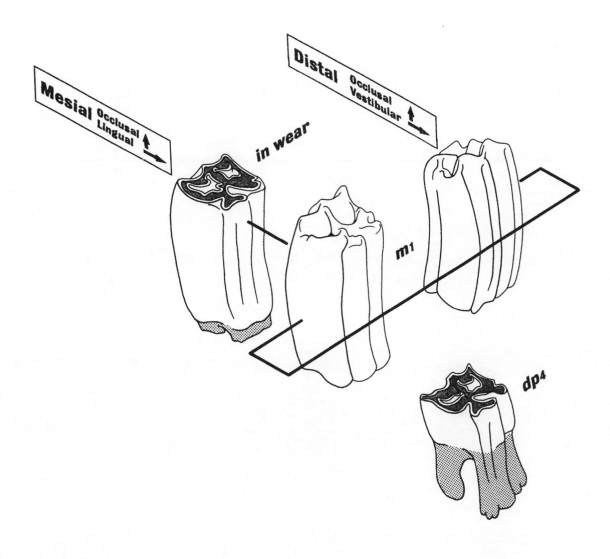

FIGURE 16. TEETH WITH AN INFUNDIBULUM (PART 3)

Horse upper left permanent first molar and deciduous fourth premolar (Group 9c in the text). Left hand diagonal row shows mesial/lingual views of the first molar (unworn and with slight wear), and the deciduous fourth premolar (moderately worn). Right hand drawing shows a distal/vestibular view of the same first molar.

Isometric projections, reproduced at approximately life-size.

9c. Bovid or cervid upper permanent first to third molars or deciduous third and fourth premolars, or lower first and second molars, horse permanent upper premolars and molars and deciduous premolars (Figures 14-16)

• **Bovid or cervid upper molars.** Crown: 'B'-shaped occlusal outline, with a double infundibulum. Four broad infolds separate five ridges down the vestibular surface. There is a lower accessory cusp just to mesial of the centre of the lingual side, in the fold of the 'B'. In the unworn state, the crown bears four main cusps, connected by ridges which surround the double infundibulum. Wear proceeds rapidly and the two parts of the infundibulum are soon isolated as islands in the 'B'-shaped occlusal surface. The third molar tapers to distal, whereas the first and second molars are much squarer. Roots: three, with the lingual being much the largest and tending to be divided into two.

• The crowns of **cervid** permanent upper molars show a prominent cervical bulge on the lingual side, so that the crown tapers to occlusal. In their unworn state, the crowns are considerably lower than those of bovids. They also have sharper vestibular ridges and the lingual accessory cusp is lower – not coming into wear until late in the functional life of the tooth. In **bovids**, the lingual accessory cusp forms a tall pillar which is soon incorporated into the worn occlusal surface.

• **Bovid or cervid deciduous upper third and fourth premolars.** Similar in form to the permanent upper first and second molars but with a lower crown, more waisted in the cervical region, and more widely spread roots.

• **Bovid or cervid lower first and second molars.** Crown: with a similarly 'B'-shaped occlusal outline to that of the upper molars, also with a double infundibulum, but the pattern is reversed so that the curves of the 'B' are this time to vestibular. The ridges are less pronounced and the occlusal outline as a whole is narrower vestibular-lingual than in the upper molars. Roots: two with oval sections.

• The crowns of **cervid** lower first and second molars show similar differences from bovids to those seen in the upper molars, although the cervical bulge is this time to vestibular and is less pronounced, so that the lower teeth can be more difficult to distinguish.

• **Horse permanent second to fourth premolars and first to third molars.** Crown: with a 'B'-shaped occlusal outline not dissimilar to that of bovids, but a much squarer, taller and more parallel-sided crown. Two infolds separate off a prominent buttress in the mesial/lingual corner. The second premolar and third molar, which form the ends of the cheek tooth row have a triangular occlusal outline (Fig. 6). Roots: three short roots which appear late in the development of the tooth and are barely distinguishable from the crown.

• **Horse deciduous second to fourth premolars.** Crown: similar in occlusal outline to the permanent premolars, but with a lower crown which is slightly waisted cervically. Roots: three more prominent and spreading roots.

9d. Bovid or cervid lower permanent third molar or deciduous fourth premolar (Fig. 15)

• **Permanent lower third molar.** Crown: similar in form to the lower second molar, with the addition of another element to distal, making it the longest tooth in the dentition. The distal element may or may not have a small infundibulum. Roots: three, arranged in a mesial-distal line, the most distal of which may not be fully separated from its neighbour.

• **Deciduous lower fourth premolar.** Crown: similar in general plan to the permanent lower first molar, but with an additional 'D'-shaped element (of similar size) at the distal end, producing a very elongated (mesial-distal) outline. The crown is markedly waisted in the cervical region, which also arches strongly between the roots. Roots: two main, widely separated roots, with one further slender root on the vestibular side.

TOOTH SOCKETS (FIGS. 17 & 18)

The form of roots is almost as distinctive as the form of the tooth crown, so that the pattern of sockets in the upper and lower jaws is useful for identification, even when the teeth themselves are missing. The figures can also be used as a guide to the number and form of tooth roots.

MANDIBLE (FIG. 19)

The main element of the mandible is the *body*, which bears the sockets for the teeth. Left and right bodies are joined at their cranial ends in the mandibular symphysis. Amongst humans and carnivores this joint is replaced by bone early in life, but in the herbivores it remains mobile throughout. A flattened area, the *ramus*, rises at the caudal end of the body. At its dorsal end the ramus expands into two processes, the *coronoid process* for attachment of the temporalis muscle and the *condyle*, which bears an articular surface for the joint with the temporal bone of the skull. The mandibular body and ramus are very common archaeological finds.

Mandibular condyle
• Rounded and roller-like; **horse, pig, human, dog, cat.**
• Saddle-shaped; **bovids, cervids.**
• Relatively narrow medial-lateral; **human, pig.**
• Relatively broad medial-lateral; **dog, cat.**
• The caudal border of the ramus is convex below the condyle in lateral view; **horse.**

Coronoid process
• Long ventral-dorsal, narrow cranial-caudal, pointed at its end and strongly curved to caudal; **bovids, cervids.**
• Moderate length, narrow cranial-caudal, rounded at the end, curved to caudal but less strongly so than in bovids and cervids; **horse.**
• Small, pointed, asymmetrical and angled to caudal; **pig.**
• Small, regular and triangular; **human.**
• Relatively long and very broad; **dog, cat;** (less long in the cat than in the dog).

Angle of body and ramus
• Bearing a prominent process; **dog, cat.**
• Strongly marked with radial ridges; **pig.**
• The border of the angle bulges strongly down to ventral; **cervids.**

VERTEBRAE (FIGS. 20-27)

Vertebrae have two major components; the *body* and the *neural arch.* Attached to the arch are five minor parts: a single *neural spine*, and left and right *articular processes* from which arise two *transverse processes.* The articular processes bear *lateral articular facets* to cranial and caudal. In some animals there is an *arch foramen* on each side of the arch (Fig. 22). The form and development of these components allow the vertebrae to be divided into five different types found in five different regions of the vertebral column:

Cervical vertebrae (Fig. 23). Prominent articular processes, neural spine and transverse processes variably developed, a large foramen (the *foramen transversarium*) penetrating the arch to medial and lateral of the body. The first cervical vertebra, the *atlas* (Figs. 20, 21), and the second cervical vertebra, or *axis* (Fig. 22), are very different from the rest of the cervicals and are amongst the most readily identifiable vertebrae.

Thoracic vertebrae (Fig. 24). Long neural spine, extra articular facets for attachment of the ribs – one at the tip of each transverse process and four at the cranial and caudal points of attachment of the arch to the body.

Lumbar vertebrae (Fig. 25). Massive body, long transverse processes, short and stout neural spine, articular processes curling over so that the facets of neighbouring vertebrae interlock.

Sacral vertebrae (Figs. 26, 27). Not unlike lumbar vertebrae initially, but fusing together in mature individuals to form a caudal tapering block which is bound into the pelvis. A broad articular surface (the *auricular area*) for the innominate bones covers the cranial-lateral margins.

Caudal vertebrae (Fig. 27). The arch and processes are reduced, with the body being the most prominent feature. The vertebrae are reduced to short rods of bone at the caudal tip.

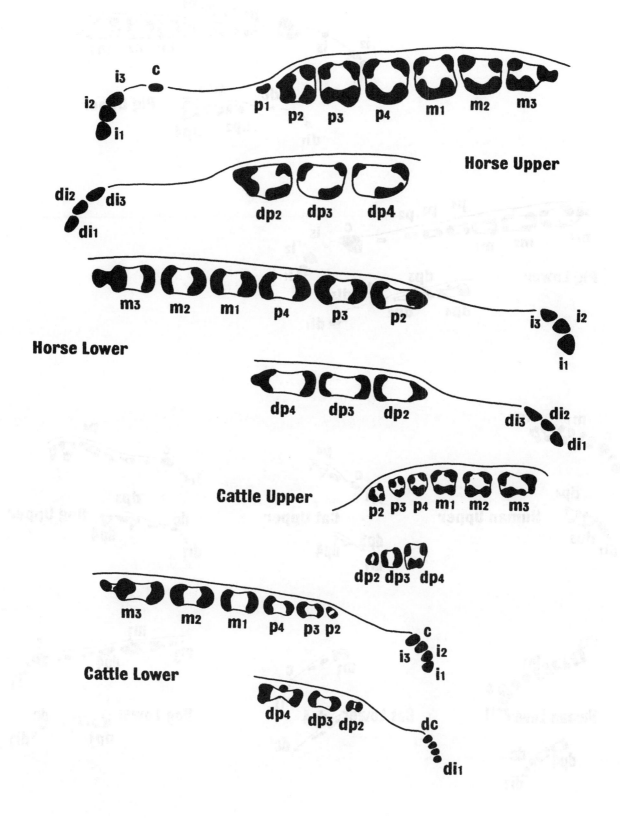

FIGURE 17. ROOT SOCKET PATTERNS (PART 1)

Outlines of root sockets for the left half of the dentition. The sockets are shaded black and, when a tooth is multi-rooted, they are joined with an outline. All are reproduced at approximately half life-size. From top to bottom: horse permanent upper, deciduous upper, permanent lower and deciduous lower dentitions, bovid (cattle) permanent upper, deciduous upper, permanent lower and deciduous lower dentitions.

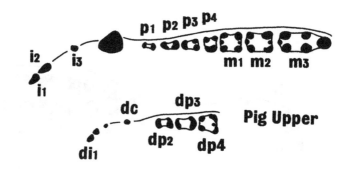

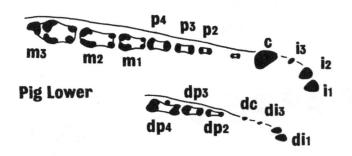

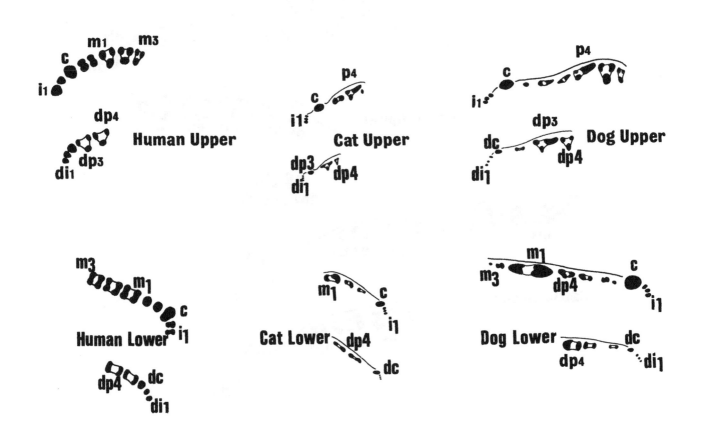

FIGURE 18. ROOT SOCKET PATTERNS (PART 2)

Outlines of root sockets for the left half of the dentition. The sockets are shaded black and, when a tooth is multi-rooted, they are joined with an outline. All are reproduced at approximately half life-size. The top four drawings are pig permanent upper, deciduous upper, permanent lower and deciduous lower dentitions. Of the three columns below: the leftmost is human permanent upper, deciduous upper, permanent lower and deciduous lower dentitions; the centre is the same for cat and the rightmost the same for dog.

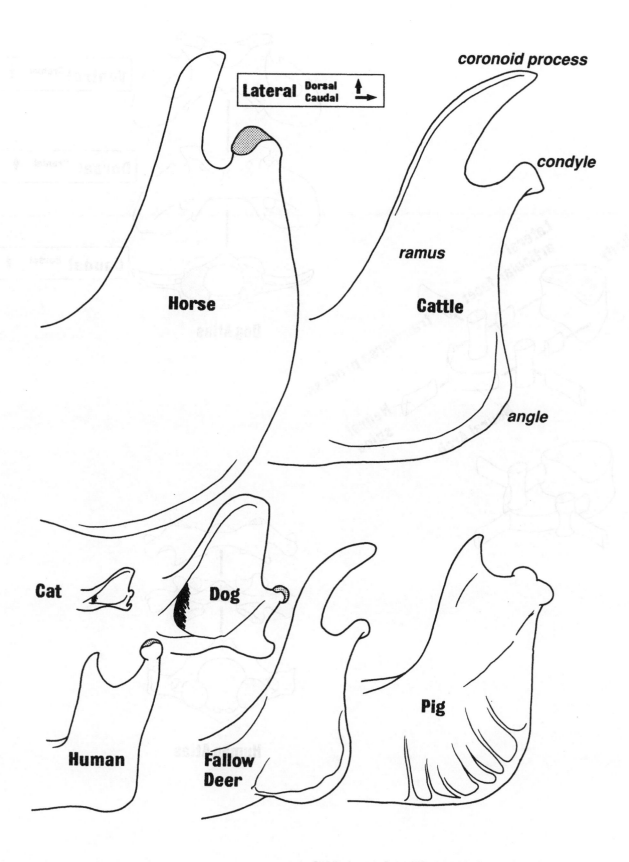

FIGURE 19. MANDIBULAR RAMUS

Lateral view of left mandibular ramus for horse, bovid (cattle), cat, dog, human, cervid (fallow deer) and pig. All approximately half life-size.

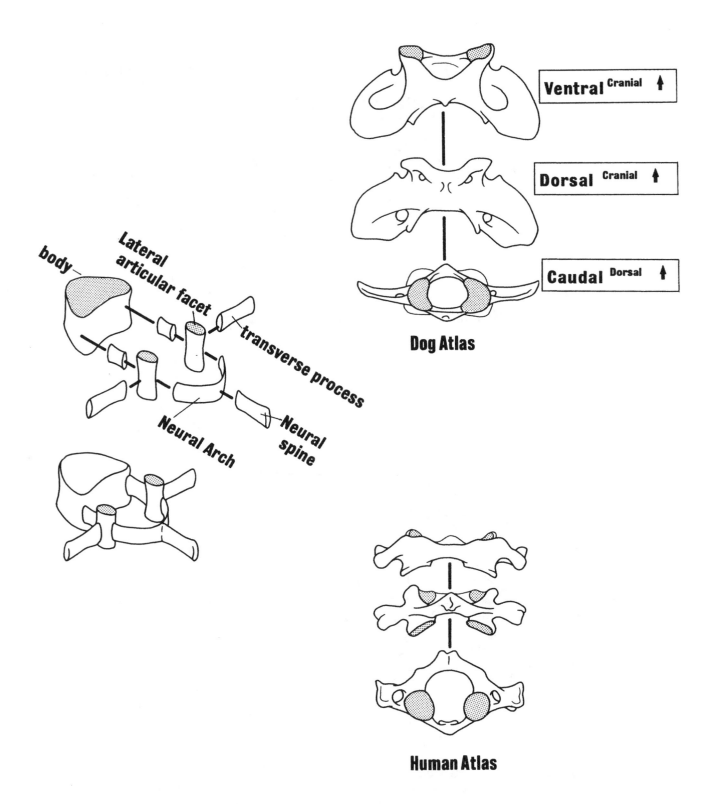

Ventral Cranial ↑

Dorsal Cranial ↑

Caudal Dorsal ↑

Dog Atlas

body

Lateral articular facet

transverse process

Neural spine

Neural Arch

Human Atlas

FIGURE 20. A GENERALISED VERTEBRA & ATLAS (PART 1)
Isometric projection of the major elements of a generalised vertebra. Not to scale. After Kapandji (1974).
Ventral, dorsal and caudal views of the atlas for dog and human. Approximately half-size.

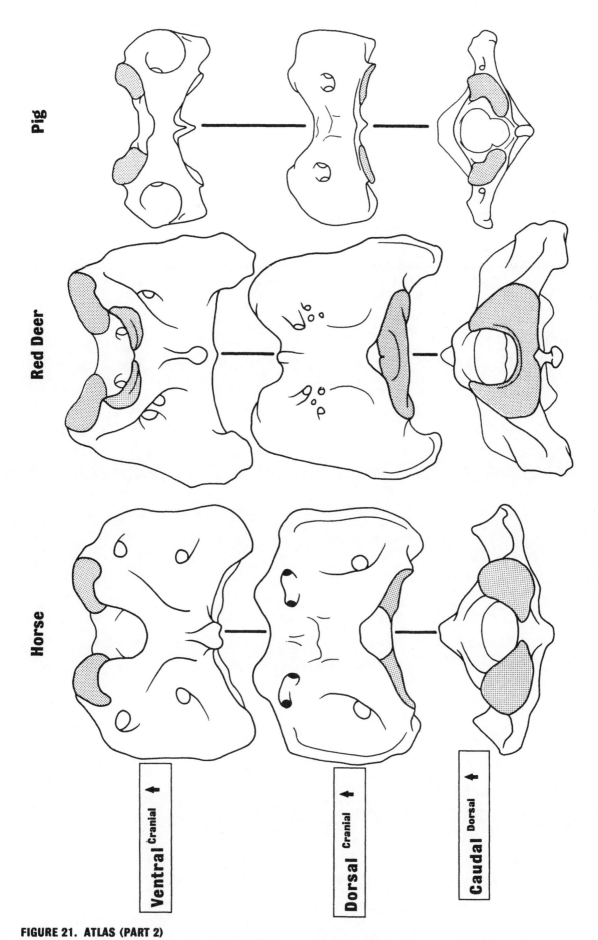

FIGURE 21. ATLAS (PART 2)

Ventral, dorsal and caudal views of the atlas for horse, cervid (red deer) and pig. Approximately half life-size.

As the different types of vertebrae approach one another, they gradually take on a form more like that of the neighbouring type, so that it is possible not only to tell which group a vertebra belongs to, but also to suggest roughly where in that group it is placed. Vertebrae from a complete skeleton can readily be assembled, jigsaw fashion, on a table top.

It is not possible in a small book to give details of all vertebrae and the following sections describe the atlas, axis, selected "typical" cervical, thoracic, lumbar and caudal vertebrae, and the overall form of the sacrum.

• Short cranial-caudal, rather square in dorsal or ventral outline, not extending much to caudal of the arch; **pig**.
• Relatively thin dorsal-ventral, spreading widely to lateral, lateral margins tapering markedly to cranial, swept markedly to caudal of the arch; **dog, cat**.

Foramen transversarium
• Hidden or absent; **bovids, cervids, pig**.
• Prominent; **horse, human, dog, cat**.

NUMBERS OF VERTEBRAE

	cervical	thoracic	lumbar	sacral	caudal
Horse	7	18	6	5	15-21
Cattle	7	13	6	5	18-20
Sheep	7	13	6-7	4	16-18
Red deer	7	13	6	4	11
Pig	7	14-15	6-7	4	20-23
Human	7	12	5	5	3-5
Dog	7	13	7	3	20-23
Cat	7	13	7	3	21-23

(Getty, 1975; Schmid, 1972; Jayne, 1898).

ATLAS (FIGS. 20 & 21)
This vertebra is the only one not to have a body. Two arches connect the articular processes, which bear large dished facets on the cranial surface and flatter facets on the caudal surface. Inside the ventral arch or wall is a special facet for the odontoid peg of the axis (below). Both ventral and dorsal arches bear tubercles at their apex. The transverse processes in most of the animals described here are large and extended into wing-like structures.

Transverse processes
• Spreading and wing-like; **horse, bovids, cervids, pig, dog, cat**.
• Short and stout, extending to lateral from a narrow ring of bone surrounding the foramen transversarium; **human**.
• Lateral margins approximately parallel, long cranial-caudal but not extending much to caudal of the arch; **horse**.
• Lateral margins somewhat converging to cranial, long cranial-caudal and extending slightly to caudal of the arch; **bovids, cervids**.

Cranial lateral joint facets
• Oval and slightly dished, widely separated by the ventral-cranial lip of the atlas, which is slightly curved to caudal; **human, dog, cat**.
• Elongated and slightly dished, and separated by a small length of the ventral-cranial margin, which runs relatively level; **pig**.
• Deeply dished, and separated by a small gap on the ventral-cranial lip of the atlas, which is prominently curved to caudal, the edge of the facets curling over the ventral-cranial lip; **bovids, cervids**.
• Deeply dished, and separated by a wide gap on the ventral side, with an even more prominent curve in the middle of the cranial-ventral lip; **horse**.

Caudal lateral joint facets
• Widely separated oval facets; **human, pig, dog, cat**.
• Separate oval facets, with their medial borders close together; **horse**.
• Facets coalesced, making a continuous crescentic articular surface; **bovids, cervids**.

25

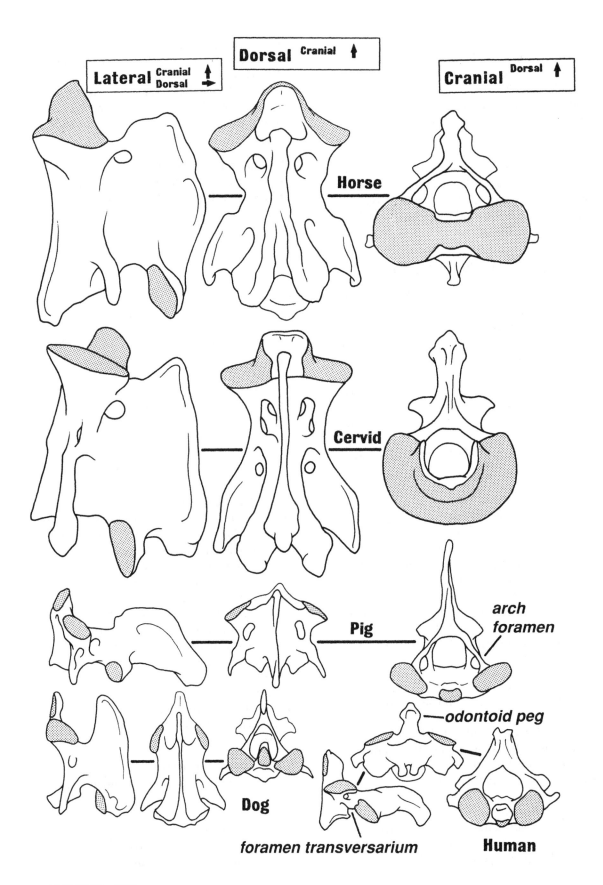

FIGURE 22. AXIS

Lateral, cranial and dorsal views of the atlas for horse, cervid (red deer), pig, dog and human.
Approximately half life-size.

AXIS (FIG. 22)

Also known as the *epistropheus*. The body bears a process (the *odontoid peg*) at its cranial end, which has a joint facet for articulation with the inside of the ventral arch of the axis. The main cranial joint facets are situated to lateral of the odontoid peg. The neural spine is prominent and highly variable in form, and bears the main caudal joint facets on its caudal end. Transverse processes are relatively small and carry the foramina transversaria. Some animals also have a prominent arch foramen at the cranial root of each side of the arch.

Body
• Relatively very short cranial-caudal; **human**.
• Relatively very long, with a pronounced ventral crest; **horse**.
• Relatively very long, with a less pronounced ventral crest; **dog**.
• Short, without a crest; **pig**.
• Intermediate length, with a moderate crest; **bovids, cervids**.

Odontoid peg
• Markedly peg-like, long cranial-caudal and with a swollen tip; **human**.
• Markedly peg-like, relatively short and stout; **pig**.
• Markedly peg-like, long and tapering to a sharp point; **dog, cat**.
• Tongue-like and long; **horse**.
• Gutter-like and relatively shorter than in the horse; **bovids, cervids**.

Cranial lateral joint facets
• Isolated and oval in outline; **human, dog, cat**.
• Bean-shaped outline and isolated from the odontoid peg; **pig**.
• Crescentic outline, confluent with the odontoid peg facet and wrapped around it like a flange, touching or almost touching at the medial-ventral facet borders; **bovids, cervids**.
• Oval outline, confluent with the odontoid peg facet, but widely spaced at the medial-ventral facet borders; **horse**.

Neural spine
• Very short cranial-caudal, stout; **human**.
• Moderate length cranial-caudal, fan-shaped in lateral view; **pig**.
• Intermediate length cranial-caudal, extending slightly beyond the cranial border of the arch, but stopping level with the caudal border; the caudal articular processes extend separately to the spine; **bovids, cervids**.
• Long cranial-caudal, extending slightly beyond the cranial border of the arch, and extending well beyond the caudal border, bearing caudal articular processes on its ventral side; **horse**.
• Long cranial-caudal, extending well beyond the cranial border of the arch, and extending well beyond the caudal border, bearing caudal articular processes on its ventral side; **dog, cat**.

Transverse processes
• Very small, not making much more than a ring of bone around the foramen transversarium; **human**.
• Similar to human, but slightly more developed; **pig**.
• Moderately prominent, wing-like and directed strongly to caudal; **horse, bovids, cervids, dog, cat**; (dog and cat transverse processes are also ventrally directed).

Foramen transversarium
• Prominent, surrounded by a narrow ring of bone; **human**.
• Prominent, making a tunnel through the root of the transverse process; **dog, cat**.
• Small to tiny and highly variable, making a tunnel through the root of the transverse process; **horse, bovids, cervids, pig**.

Arch foramen
• Normally absent; **human, dog, cat**.
• Prominent and close to the cranial edge of the arch; **horse**.
• Prominent and further from the cranial edge of the arch; **bovids, cervids, pig**.

CERVICAL VERTEBRAE OTHER THAN ATLAS AND AXIS (FIG. 23)

The remaining cervical vertebrae vary greatly down their series from third to seventh, but there is not space to describe the differences in this book. The sixth has been chosen as a representative. All the animals described here have a pronounced foramen transversarium in this vertebra.

Arch foramina
• **Pig**; cervical vertebrae have arch foramina.

Cranial and caudal articulation of body
• Cranial articular surface markedly domed into a "head", the caudal surface is correspondingly deeply dished, with flared borders; **horse**.
• Similar cranial doming and caudal dishing, but less pronounced; **bovids, cervids, pig**.
• Cranial surface somewhat bulging, caudal surface slightly dished; **dog, cat**.
• Cranial and caudal articular surfaces approximately flat, with additional facets for synovial joints on their lateral borders; **human**.

Neural spine
• Small, reduced in sixth cervical vertebra to a low ridge and small cranial tubercle; **horse**.
• Short and angled to caudal, often with two tubercles at its tip; **human**.
• Considerably more prominent than in horse or human, stout and angled towards cranial; **bovids, cervids**.
• Similarly prominent and cranially angled, but longer, more slender and more pointed; **pig**.
• Prominent, pointed, cranially angled, but relatively short; **dog, cat**.

Transverse process
• Represented by small tubercles, arising from a thin ring of bone around a large foramen transversarium; **human**.
• Prominent and with two elements, a ventral plate and a lateral tubercle; **horse, bovids, cervids, pig, dog, cat**.

• Ventral plate especially prominent, thin and with a curving ventral border which extends symmetrically to cranial and caudal of the body; **pig**.
• Ventral plate prominent and robust, with a relatively straight and swollen ventral border, extending asymmetrically with a slight tilt to cranial; **bovids, cervids**.
• Ventral plate present, but less prominent, thinner and less extensive to caudal; **dog, cat**.
• Ventral plate present and robust, but not so prominent, especially at its caudal end, extending only a little way to ventral of the lateral tubercle; **horse**.

THORACIC VERTEBRAE (FIG. 24)
The sixth thoracic vertebra has been chosen to represent the group as a whole, but the more cranial and more caudal thoracic vertebrae may depart quite markedly in form.

Arch foramina
• **Pig** thoracic vertebrae have arch foramina.

Neural spine
• Flattened, blade-like, with a sharp edge to caudal, a more bulging and blunter edge to cranial, and a bulbous roughened dorsal termination; **horse**.
• Similar, but with less bulging section, caudal edge sharper and with a less bulbous dorsal termination; **bovids, cervids**.
• Similar in size of vertebral body to large cattle, but with a markedly longer spine; **bison**.
• Broad and rounded in section; **pig**.
• Triangular in section; **human, dog, cat**.

Lateral articular facets
• Far apart; **human**.
• Close together; **horse, bovids, cervids, pig, dog, cat**.

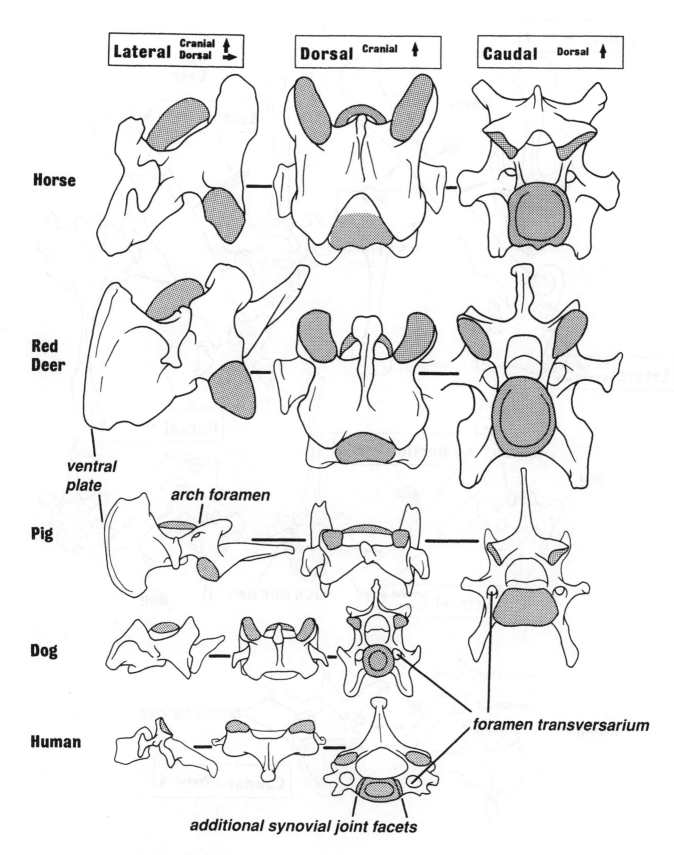

| Lateral Cranial↑ Dorsal→ | Dorsal Cranial↑ | Caudal Dorsal↑ |

Horse

Red Deer

ventral plate

arch foramen

Pig

Dog

foramen transversarium

Human

additional synovial joint facets

FIGURE 23. CERVICAL VERTEBRAE
Lateral, dorsal and caudal views of the sixth cervical vertebra for horse, cervid (red deer), pig, dog and human. Approximately half life-size.

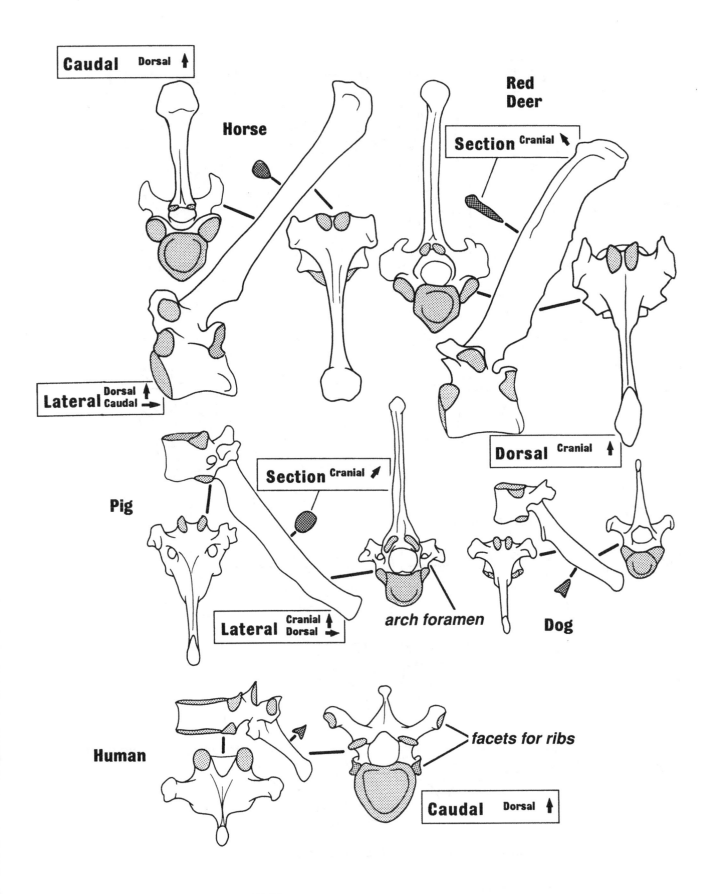

Caudal Dorsal ↑

Horse

Red Deer

Section Cranial ↑

Lateral Dorsal ↑ Caudal →

Dorsal Cranial ↑

Pig

Section Cranial ↗

arch foramen

Lateral Cranial ↑ Dorsal →

Dog

Human

facets for ribs

Caudal Dorsal ↑

FIGURE 24. THORACIC VERTEBRAE
Caudal, lateral and dorsal views, with a mid-section of the neural spine, for the sixth thoracic vertebra of horse, cervid (red deer), pig, dog and human. Approximately half life-size.

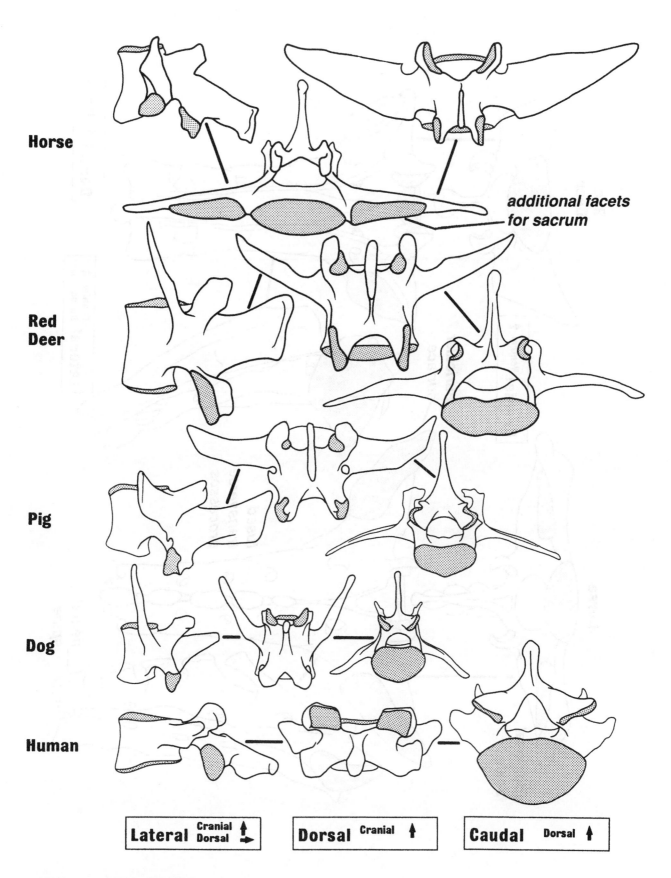

Horse

Red Deer

Pig

Dog

Human

additional facets for sacrum

Lateral	Cranial ↑ Dorsal →
Dorsal	Cranial ↑
Caudal	Dorsal ↑

FIGURE 25. LUMBAR VERTEBRAE

Lateral, caudal and dorsal views of horse fifth lumbar vertebra, cervid (red deer), pig and dog sixth lumbar vertebra, and human fifth lumbar vertebra. Approximately half life-size.

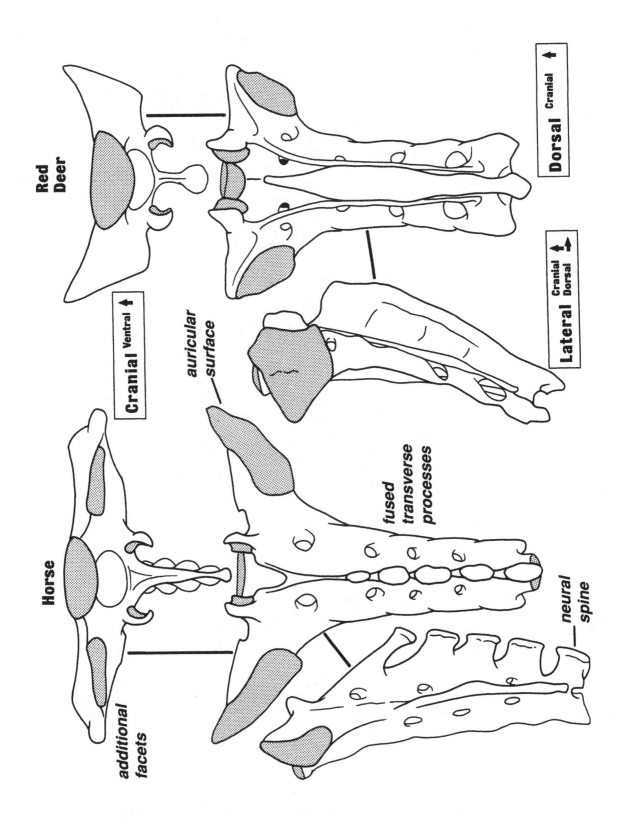

Red Deer

Horse

Dorsal ^{Cranial}

Lateral ^{Cranial} ^{Dorsal}

Cranial ^{Ventral}

auricular surface

fused transverse processes

neural spine

additional facets

FIGURE 26. SACRUM (PART 1)

Cranial, dorsal and lateral views of the sacrum for horse and cervid (red deer). Approximately half life-size.

LUMBAR VERTEBRAE (FIG. 25)

There are again variations along the lumbar series, but the lowest of them (fifth, sixth or seventh) has been chosen for the description here because it shows the most clear-cut differences between animals.

Arch foramina
• **Pig**; lumbar vertebrae have arch foramina.

Neural spine
• Markedly broad cranial-caudal relative to length dorsal-ventral, stout and square in lateral outline, somewhat tilted to cranial; **horse, bovids, cervids, pig**.
• Similar, but relatively even shorter dorsal-ventral, and somewhat tilted to caudal; **human**.
• Narrower cranial-caudal relative to length dorsal-ventral, less stout than the above, triangular in lateral outline, tilted to cranial; **dog, cat**.

Transverse process
• Long medial-lateral, flattened in section and blade-like, broad cranial-caudal, extending horizontally to lateral (or with a slight angle to ventral), somewhat curved to cranial along its length; **horse**.
• Similar, but more markedly curved to cranial; **bovids, cervids, pig**; (cattle and pig markedly robust).
• Long medial-lateral, flattened in section, tapering in outline, somewhat angled to ventral, strongly angled to cranial; **dog, cat**.
• Short medial-lateral, stoutly built, slightly angled to dorsal and to cranial; **human**.
• Bearing large triangular facets for articulation with the sacrum; **horse**.

SACRUM (FIGS. 26 & 27)

Number of fused vertebrae
• Five sacral vertebrae; **horse, bovids, cervids, human**.
• Four sacral vertebrae (but first caudal may also fuse); **pig**.
• Three sacral vertebrae; **dog, cat**.

Neural spines
• Reduced to low bumps, unfused and effectively missing; **pig**.
• Short dorsal-ventral, stout, only partly fused, the dorsal tips not linked by a high crest; **human**.
• Similar, but smaller, flatter, thinner and less stout; **dog, cat**.
• Long and stout, partly fused together, with a discontinuous crest joining the dorsal tubercles; **horse**.
• Slightly shorter, but still stout, fully fused together, with a continuous bar of bone joining the dorsal tubercles; **bovids, cervids**.

Auricular area
• Triangular outline, markedly spreading to lateral; **horse**.
• Squarer outline, less spreading to lateral; **bovids, cervids, pig**.
• Bean-shaped outline, spreading only a little to lateral; **human, dog, cat**.

Transverse processes
• Auricular surfaces borne by the transverse processes of the first sacral vertebra, the rest being fused into a stout plate which tapers slightly medial-lateral to caudal; **horse**.
• Similar, but with the plate not tapering to caudal; **bovids, cervids**.
• Similar, but with the plate much narrower medial-lateral; **pig**.
• Auricular surfaces borne by the transverse processes of the first sacral vertebra, the rest being fused into irregular masses which taper to caudal; **dog, cat**.
• Similar, but the auricular surfaces borne by the transverse processes of the first, second and often the third sacral vertebrae; **human**.

Cranial articulation
• Large additional triangular facets for articulation with the fifth lumbar vertebra (above); **horse**.
• Processes bearing the lateral facets lie directly on the dorsal surface of the transverse processes, with the auricular area aligned vertically dorsal-ventral, giving a compact and square cranial outline to the sacrum; **dog**.
• Lateral facet processes rising up separately to the transverse processes, with the auricular area more sloping; **cat**.

CAUDAL VERTEBRAE (FIG. 27)

The second caudal vertebra has been taken here as the model. The remainder are markedly reduced and show few distinctive features.

Neural spine

• Broad medial-lateral and stout, somewhat swept back to caudal; **horse**.
• More prominent and markedly swept back to caudal; **bovids, cervids, pig, dog, cat**.
• Arch and neural spine missing; **human**.

Transverse processes

• Relatively thick lobes, bulging to caudal; **horse**.
• Thin, triangular plates, angled to caudal; **bovids, cervids**.
• Thin, square plates, not angled to caudal; **pig**.
• Narrow, finger-like, markedly angled to caudal; **dog, cat**.
• Small, broad lobes angled to cranial; **human**.

RIBS (FIG. 28)

Ribs, or fragments of them, are some of the commonest finds but are also some of the most difficult to identify. They may be divided into two main components – the long, curving *body* and the complex of joints and tubercles which make up the proximal articulation. This comprises a proximal *head* bearing twin joint facets (to articulate with pairs of facets on the bodies of neighbouring thoracic vertebrae), a *neck* connecting it with the body, and a *tubercle* which bears a small facet (to articulate with the facet on the transverse process of a thoracic vertebra).

Section of body

There is a great deal of variation from cranial to caudal in the rib row and the section of the body is not a very reliable guide to identification. The descriptions below are for ribs about midway along in the series.
• Large and robust, with an oval section; **horse**.

• Large and robust, with an aerofoil-like section produced by a 'flange' along the caudal edge; **large bovids**.
• Intermediate size, with aerofoil section; **cervids, small bovids** ('flange' usually less pronounced in cervids).
• Intermediate size, plump and rounded in section; **pig**.
• Intermediate size, oval in section, sometimes with a ridge along the cranial and caudal sides; **human**.
• Small size; **dog, cat**. Various sections from rounded to flattened. Cat very much smaller than dog.

Proximal articulation

• Two clearly separated oval facets on the head; **horse, bovids, cervids, pig**.
• Concave, oval facet on the tubercle; **horse, bovid, cervid**.
• Convex, round facet on the tubercle, with relatively large and rounded head facets; **pig**.
• The neck is set at a slightly smaller angle to the body in **cattle** than in **horse**.
• Twin head facets combined into a single domed facet; **dog, cat**.
• Twin head facets combined into a single saddle-shaped facet; **human**.

Costal cartilages

In most mammals, the distal end of the ribs is joined to the sternum, or breastbone, by cartilagenous processes. These are, however, gradually replaced by bone with increasing maturity and "ossified" costal cartilages are quite common finds. They are rounded, cigar-shaped bodies of highly porous bone with a coarsely sponge-like texture.

SCAPULA (FIGS. 29 & 30)

The scapula is distinguished by its thin, triangular *blade* which is, unfortunately, often damaged in archaeological specimens. The blade has cranial and caudal *borders* which may or may not be thickened, and a *spine* rising above its dorsal surface. At its distal end, the spine is often swollen into a process called the *acromion* (to which the deltoid muscle is attached). A dished articular

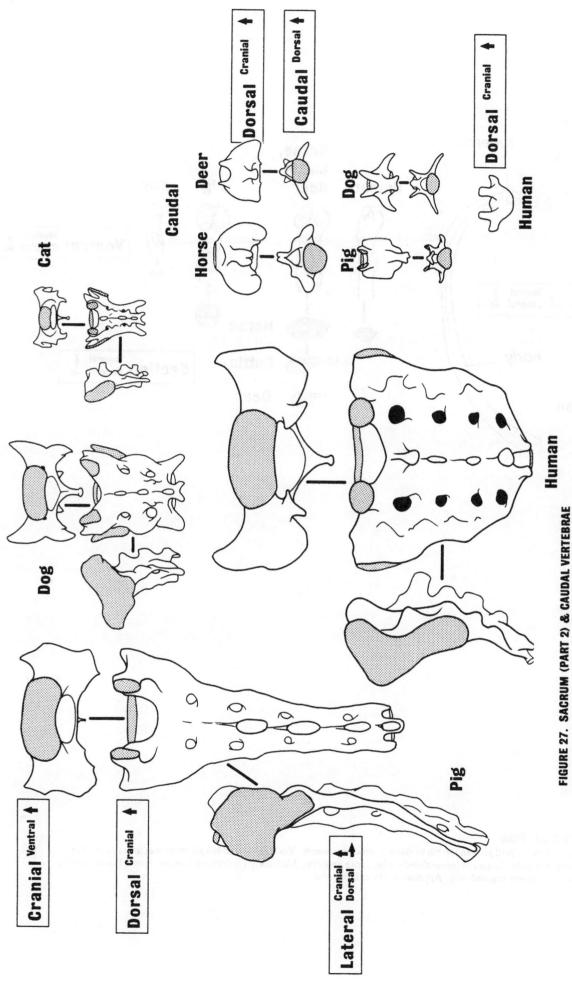

FIGURE 27. SACRUM (PART 2) & CAUDAL VERTEBRAE

Cranial, dorsal and lateral views of the sacrum for pig, human, dog and cat.
Dorsal and caudal views of second caudal vertebra for horse, cervid (red deer), pig and dog. Dorsal view of first
caudal vertebra for human.
Approximately half life-size.

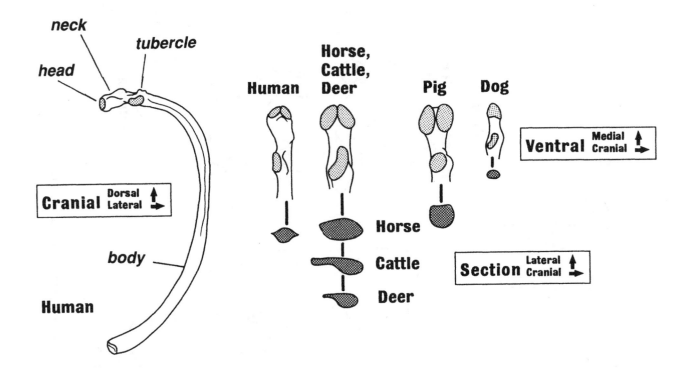

FIGURE 28. RIBS

Cranial view of fifth human left rib to show main components. Ventral view of left rib proximal articulation for human, horse-bovid-cervid (generalised outline), pig and dog. Mid-body sections for human, horse, bovid (cattle), cervid (red deer), pig and dog. Approximately half life-size.

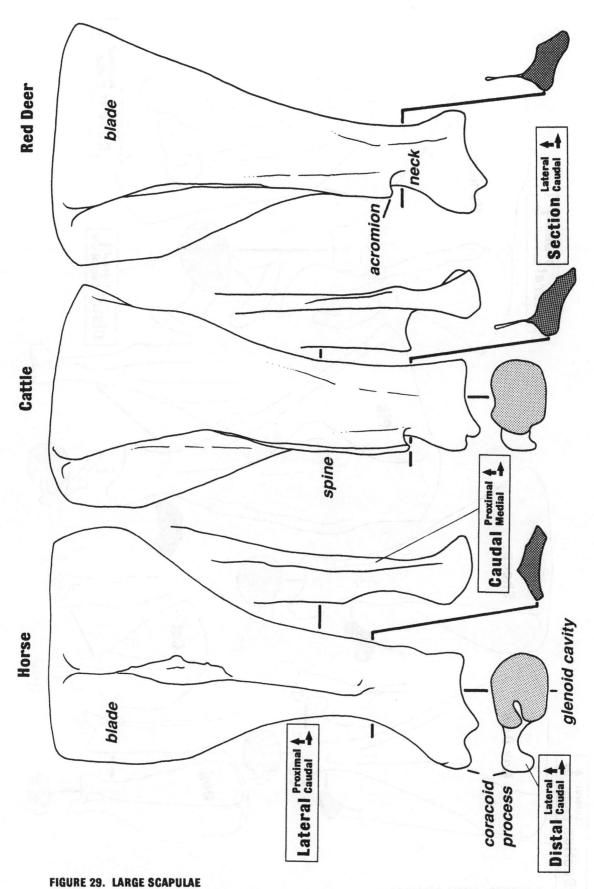

FIGURE 29. LARGE SCAPULAE

Lateral, caudal and distal views, with a section at the neck, for the left scapula of horse, large bovid (cattle) and large cervid (red deer). Approximately half life-size.

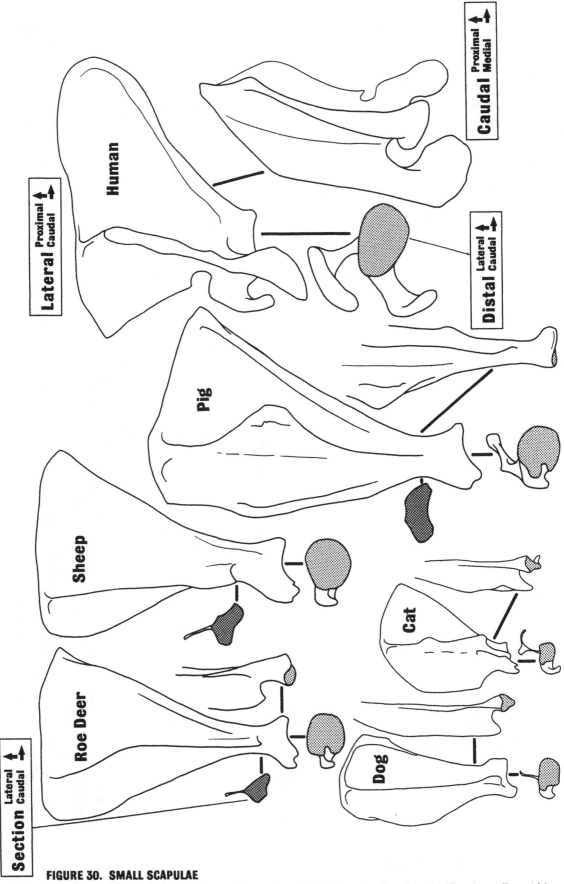

FIGURE 30. SMALL SCAPULAE
Lateral, caudal and distal views, and a neck section, for the left scapula of small bovid (sheep), small cervid (roe deer), dog, cat, pig and human. Approximately half life-size.

surface, the *glenoid cavity* sits at the distal apex of the blade, to which it is connected by a narrow neck. The *coracoid process* extends out from the cranial side of the neck and glenoid cavity.

Blade outline
• Large, regular isosceles triangle; **horse, large bovids, large cervids.**
• Intermediate-sized, regular isosceles triangle; **fallow deer.**
• Intermediate-sized, broad, irregular triangle; **human.**
• Small, regular isosceles triangle; **small bovid, roe deer.**
• Small to intermediate-sized, broad triangle; **pig.**
• Very small to small, narrow triangle, with a curved cranial border; **dog, cat.**

Border thickening
• Only the caudal border is thickened; **horse, bovids, cervids, human, dog, cat.**
• Both the caudal and the cranial borders are markedly thickened; **pig.**

Neck
• Long (proximal-distal), broad (cranial-caudal) and markedly flattened (medial-lateral); **horse.**
• Long (proximal-distal), narrower (cranial-caudal) than in horse, but stoutly built and less markedly flattened (medial-lateral); **bovids.**
• Similar to bovids, but narrower (cranial-caudal) relative to length (proximal-distal) and displaying marked waisting; **cervids** (waisting least marked in **caribou**).
• Shorter (proximal-distal) and more stoutly built, oval cross-section; **pig.**
• Shorter still (proximal-distal), relative to the blade, and only just recognisable as a separate neck; **dog, cat.**
• Even shorter (proximal-distal), and not really recognisable as a separate neck; **human.**

Spine
• Rising gradually from the proximal border of the blade and maintaining a high, relatively straight edge roughly at right-angles to the blade, with no or only a slight tuberosity along its crest; **bovids, cervids.**

• Rising gradually from the proximal border of the blade to form a relatively lower edge, with a prominent tuberosity along its crest and sloping gradually down to merge with the neck of the scapula; **horse.**
• Spine triangular in outline, rising gradually from the proximal border of the blade to an apex which is bent sharply to overhang the caudal part of the blade, and then sloping gradually down to merge with the neck; **pig.**
• Rising sharply from the proximal border of the blade to a high edge, much swollen by a tuberosity, and continued at a high level above the neck of the scapula; **human.**
• Rising in a smooth curve from the proximal border of the blade to a high edge, not markedly developing a tuberosity, but running on straight, at right-angles to the blade and at a high level, above the neck of the scapula; **dog.**
• Rising in a smooth curve from the proximal border of the blade to a high edge, curving over in its distal two thirds to overhang the caudal blade surface, and maintained at a high level above the neck of the scapula; **cat.**

Acromion
• Effectively absent; **horse, pig.**
• Represented by a small tuberosity at the crest of the sharply angled distal end of the spine, rising above the neck of the scapula; **bovids, cervids.**
• Pronounced and expanded into a spatulate tuberosity, overhanging the neck and the glenoid cavity by a wide margin; **human.**
• Expanded into a prominent wedge- or tongue-shaped tuberosity overhanging the neck of the scapula; **dog, cat** (a little more prominent in the cat).

Coracoid process
• Broad (medial-lateral), low, and little separated from the cranial border of the glenoid cavity; **pig.**
• Moderately developed, and clearly distinguished from the cranial border of the glenoid cavity, but not prominent; **bovids, cervids, dog.**
• More strongly developed and considerably more extensive cranially than in pig, dog, bovids or cervids; **horse.**
• Strongly developed and expanded medially into a finger-like projection; **human, cat.**

HUMERUS (FIGS. 31-34)

The most proximal long bone of the forelimb, the humerus has a swollen, rounded proximal articulation, its *head*, for the shoulder joint with the scapula. Adjacent to this are two tuberosities, the *greater tubercle* (for attachment of the supra- and infra-spinatus muscles) and the *lesser tubercle* (for attachment of the subscapularis muscle), between which lies the *intertubercular groove* (for a tendon of the biceps muscle). On the lateral surface of the mid-shaft is a prominent tuberosity for the deltoid muscle, connected to the greater tubercle by a ridge. The distal end of the bone is expanded and holds a pulley-like articular surface for the elbow joint, the *trochlea*, to the lateral end of which is joined a further surface, the *capitulum*. On the caudal side of these elements is a deep pit, the *olecranon fossa*, with a similar pit to cranial, called the *coronoid fossa*. The distal end of the humerus divides either side of the these fossae into medial and lateral *epicondylar areas*. In some animals, clear processes extend out from these, the *medial epicondyle* and the *lateral epicondyle*. Some animals also bear a sharp crest on the caudal-lateral corner of the lateral epicondylar area.

Overall size and proportions
• Large and robustly built; **horse, large bovid, moose** (the shaft of the moose humerus is markedly slender for its length).
• As long (proximal-distal) as above, but with the shaft relatively much more slender; **human**.
• Intermediate in length (proximal-distal); **caribou, fallow deer, red deer, large pig** (the shaft of the pig humerus is stouter relative to length than the others).
• Small; **small bovid, roe deer, small pig, large dog** (roe and dog humerus are markedly slender relative to their length).
• Very small; **small dog, cat** (cat humerus generally smaller than the smallest dog humerus).

Mid-shaft section
• Round section; **horse, bovids** (slight asymmetry due to caudal-lateral ridge allows the left and right humerus to be distinguished even from shaft fragments).

• Similar shape of section to horse and bovids, but small relative to the shaft length (proximal-distal); **cervids**.
• Similar section to bovids, but larger relative to shaft length and more oval in outline; **pig**.
• Markedly regular oval section, deeper cranial-caudal than pig; **dog, cat**.
• Square section; **human**.

Deltoid tuberosity and ridge to greater tuberosity
Development varies with maturity – the comments below apply only to mature animals.
• Very prominent, extending along the proximal third of the shaft; **horse**.
• Moderately prominent, extending along the proximal third of the shaft; **bovids**.
• Somewhat less strongly developed than in bovids, extending along the proximal quarter to proximal third of the shaft; **cervids**.
• Moderately prominent, confined to the proximal quarter of the shaft; **pig**.
• Represented by a broad, roughened area, not especially prominent but dictating the form of the proximal half of the shaft; **dog**.
• Similar to dog, but shorter (extending over the proximal third to half of the shaft); **cat**.
• Marked only by a slight roughening and swelling, extending along the proximal half of the shaft (the shaft is markedly straight along its length); **human**.

Proximal end – the head articulation
• Swollen and spherical in form, dominating the proximal end of the humerus; **human**.
• Broad (medial-lateral) and with a medial lip, somewhat like the cushion of a stuff-over chair; **horse, bovids, cervids, pig, dog, cat**.
• Cushion-like but less broad medial-lateral than in horse, bovids and cervids, markedly bulging and with a more oval proximal outline; **pig**.
• Similar to pig, but less bulging; **dog, cat** (cat slightly more bulging than dog).

Proximal end – greater and lesser tubercles
• The greater tubercle is very much larger than the lesser tubercle and is divided into a rounded flattened lobe to lateral, with a more pointed medial portion, the tip of which arches over the intertubercular groove; the lesser tubercle is developed into a lower

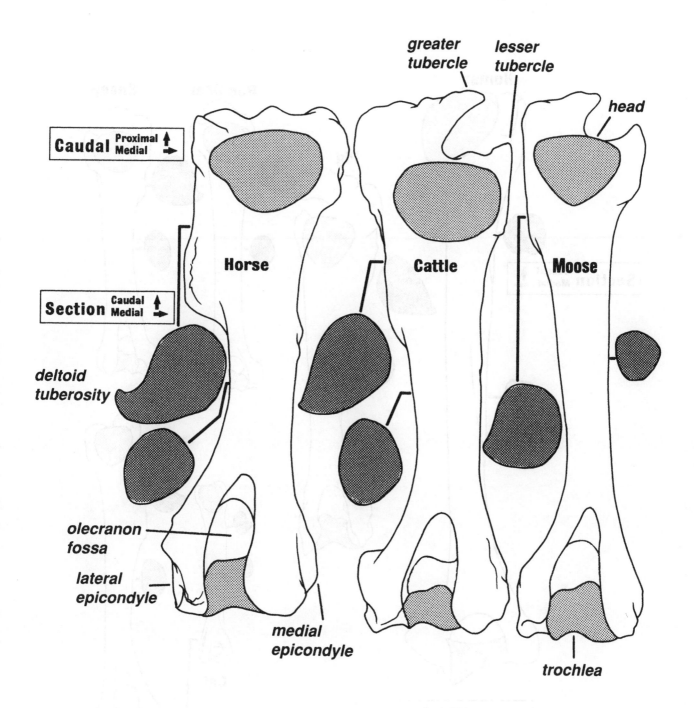

greater tubercle

lesser tubercle

head

Caudal Proximal Medial

Horse

Cattle

Moose

Section Caudal Medial

deltoid tuberosity

olecranon fossa

lateral epicondyle

medial epicondyle

trochlea

FIGURE 31. LARGE HUMERUS
Caudal views, with mid- and proximal shaft sections, for the left humerus of horse, large bovid (cattle) and large cervid (moose). Approximately half life-size.

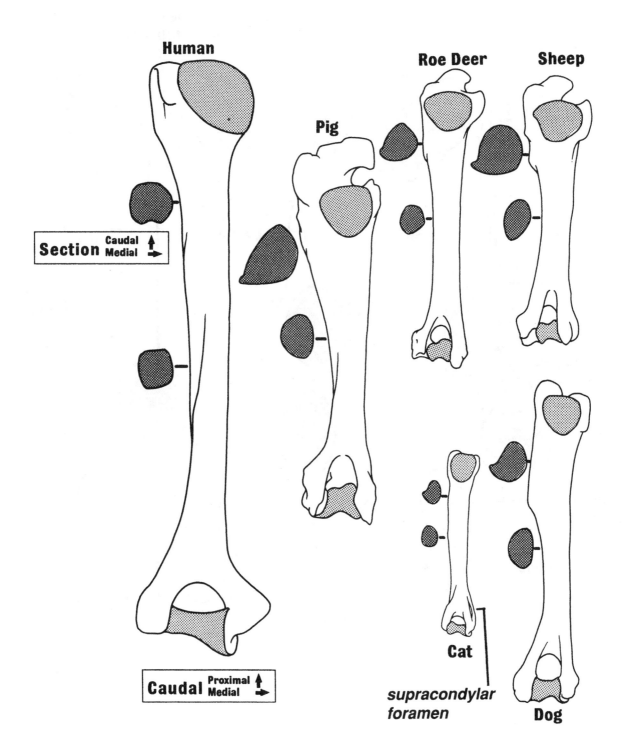

Human

Pig

Roe Deer

Sheep

Section | Caudal ↑ Medial →

Caudal | Proximal ↑ Medial →

Cat

supracondylar foramen

Dog

FIGURE 32. SMALL HUMERUS

Caudal views, with mid- and proximal shaft sections, for the left humerus of human, pig, small cervid (roe deer), small bovid (sheep), dog and cat. Approximately half life-size.

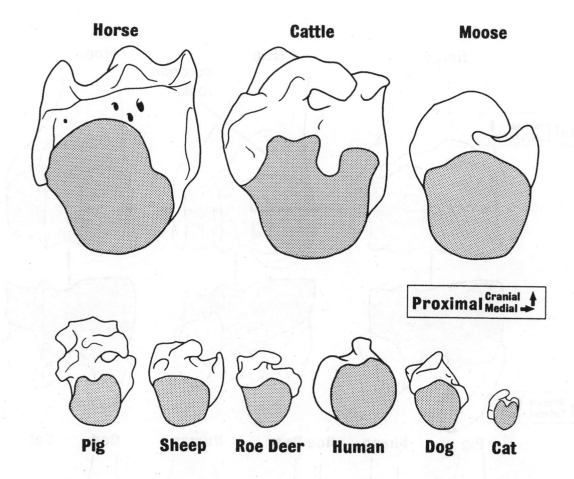

Horse   **Cattle**   **Moose**

Proximal Cranial Medial

Pig   **Sheep**   **Roe Deer**   **Human**   **Dog**   **Cat**

FIGURE 33. HUMERUS PROXIMAL ARTICULATION

Proximal views for the left humerus of horse, large bovid (cattle), large cervid (moose), pig, small bovid (sheep), small cervid (roe deer), human, dog and cat. Approximately half life-size.

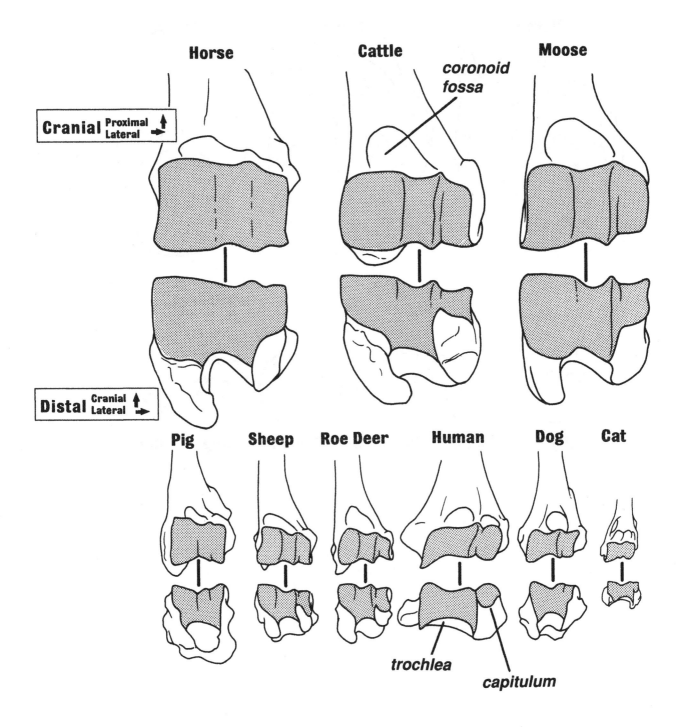

FIGURE 34. HUMERUS DISTAL ARTICULATION

Cranial and distal views of distal articulation for the left humerus of horse, large bovid (cattle), large cervid (moose), pig, small bovid (sheep), small cervid (roe deer), human, dog and cat. Approximately half life-size.

crest; **bovids** and **cervids** (the lateral lobe of the greater tubercle is less strongly developed in many cervids).

• The greater tubercle is massively developed – the lateral lobe is more clearly separated and the medial lobe is less pointed than in bovids, but it still heavily overhangs the intertubercular groove; the lesser tubercle is more prominent than in bovids; **pig**.

• Neither the greater nor the lesser tubercle is very prominent, with an intermediate tubercle between them; **horse**.

• The lesser tubercle is little developed and does not rise to proximal above the head; the greater tubercle is moderately prominent, just rising above the head and forming a bar-like lobe when seen in proximal view; **dog**, **cat**.

• The greater and lesser tubercles are not prominent and do not rise above the head, but they do have a marked intertubercular groove between them, on the cranial-lateral side of the bone; **human**.

Distal end – trochlea and capitulum

• The capitulum has a markedly domed articular surface, clearly distinguishable from the trochlea, the articulation as a whole is greatly expanded (medial-lateral) relative to the shaft, with the medial epicondyle very prominently developed, but the articulation is shallow cranial-caudal; **human**.

• The capitulum and trochlea are combined into one pulley-like articulation, with a groove (variably developed) defining the boundary between them, the capitulum is confined to a band around the lateral end of the articulation, and the trochlea is divided by a groove into a broad medial portion and a narrow lateral ridge; **horse**, **bovids**, **cervids**, **pig**, **dog**, **cat**.

• The articulation is narrow medial-lateral, with ridges and grooves very sharply defined; **cervids**.

• The articulation is broader medial-lateral than in cervids, with the ridges and grooves still clear, but less sharply defined; **bovids**.

• The articulation has a similar general form to the bovids, but the ridges and grooves are seen only as shallow undulations; **horse**.

• The articulation is relatively narrow medial-lateral and deep cranial-caudal, the groove demarcating the capitulum is moderately developed, but the particularly deep groove in the trochlea gives the articulation as a whole the appearance of two opposing cones; **pig**.

• The capitulum area is tiny and separated from the trochlea by a scarcely traceable groove, whereas the trochlea has a deep groove and a sharply prominent medial lip; **dog**, **cat**.

Distal end – olecranon fossa and coronoid fossa

• The olecranon fossa is markedly deep and enclosed by a prominent development of the epicondylar areas; **horse**, **bovids**, **cervids**, **pig**, **dog**, **cat**.

• The olecranon fossa is not notably deep, because the epicondylar areas are less strongly bulging to caudal; **human**.

• The coronoid fossa is prominent, but occupies a narrow zone just to proximal of the trochlea; **horse**, **bovids**, **cervids**, **pig**.

• The coronoid fossa is prominent and occupies a large area to proximal of the trochlea; **dog**, **cat** (the fossa is less extensive in the cat than in the dog).

• A perforation often connects the olecranon and coronoid fossae; **dog**, **cat**, sometimes **pig**.

Distal end – form of epicondylar areas

• The epicondyles are broad and spreading, the medial epicondyle bears a prominent tubercle to medial and the lateral epicondyle bears a smaller tubercle to lateral; **dog**, **cat**, **human** (human humerus epicondyles are more developed in this way).

• A slot-like *supracondylar foramen* perforates the medial epicondyle, whilst a prominent ridge runs along the caudal surface of the lateral epicondyle; **cat**.

• Neither epicondyle spreads much medial-lateral relative to the articular area, and modest tubercles only are present on their medial and lateral sides (the lateral tubercle is the more prominent of the two); **horse**, **bovids**, **cervids**, **pig**.

• A marked ridge runs along the lateral edge of the caudal surface of the lateral epicondyle; **horse**.

• A similarly prominent ridge to horse runs along the lateral epicondyle, but the lateral tubercle of the epicondyle is slightly more prominent than in horse, bovids or cervids; **pig**.

RADIUS–ULNA (FIGS. 35-37)

The ulna is a long bone with a tapering shaft, bearing at its distal end in some animals a somewhat swollen articulation, called in humans the *ulna head* (Fig. 37). At the proximal end, the articulation for the trochlea of the humerus is carried in a deep notch, the *trochlear notch* (Fig. 36). To proximal of this extends a lever-like process for attachment of the triceps muscle, called the *olecranon*. In some animals, the distal end of the trochlear notch is extended into a prominent *coronoid process*. The radius runs alongside the ulna. In humans, the radius is able to rotate around the ulna, allowing the hand to be twisted from side-to-side. With the thumb out to lateral, the radius lies parallel to, and to lateral of, the ulna. With the thumb in to medial, the shaft of the radius crosses over to cranial of the ulna, so that the radius remains to lateral of the ulna at its proximal end, but to cranial and medial of the ulna at its distal end. In quadrupeds, the radius moves much less and is held in the crossed-over (thumb to medial) position. In most hoofed mammals (ungulates), the two bones are fused in this position and the ulna's shaft and distal end are very greatly reduced. The radius has a swollen proximal articular area which is called the *radius head*, whose form varies greatly. In animals with a separate radius and ulna, there are facets for articulation between the two bones on both the radius head and the trochlear notch area of the ulna. The distal end of the radius bears a complex surface for the joints with the carpal bones, which again varies widely in form. Animals with separated radius and ulna also have surfaces on the head of the ulna and the lateral surface of the distal radius, for articulation between them.

Form of ulna
• Ulna and radius remain separate throughout life; **pig, human, dog, cat.**
• Ulna and radius are separate in immature animals, but the ulna shaft is much reduced in breadth and fuses to the radius with maturity; **bovids, cervids.**
• The ulna and radius are separate in immature animals, fusing together with maturity, but the ulna is so reduced in length that only the proximal end remains; **horse.**

Size and shape of combined radius and ulna
• Long (proximal-distal) and robust; **horse, large bovids.**
• As long as horse and large bovids, often longer, but the shaft is only moderately robust; **moose.**
• As long as large bovids, but the shaft is markedly slender; **red deer, caribou.**
• Of similar form to red deer and caribou, but slightly smaller; **fallow deer, roe deer** (the bones in roe deer are a little shorter than in fallow).
• Considerably shorter than fallow and roe deer, and more robust; **small bovids.**

Size and shape of separate radius and ulna
• Both bones as long (proximal-distal) as in the large bovids above, but very slender for their length; **human.**
• Both bones approximately the same length as for the combined bones of small bovids, but markedly stout; **pig.**
• Both bones as long as the combined bones of fallow deer; or as short as small bovid, and moderately robust; **dog.**
• Shorter than in dog, and more slender (almost stick-like); **cat.**

Shaft section
• Massively bulging radius section, with a fused ulna shaft remnant on the caudal-lateral corner in mature animals; **horse, bovids, cervids.**
• The radius section is deep cranial-caudal, symmetrical and with the caudal side rough and flattened; the ulna shaft remnant is tiny, if it extends to the mid-shaft at all; **horse.**
• The radius is less deep cranial-caudal but still robust and moderately symmetrical, with a somewhat flattened caudal side; there is a larger ulna shaft remnant than in horse; **bovids.**
• The radius shaft is rather narrower medial-lateral than in bovids, its section is markedly asymmetrical with, in some animals, a dished caudal side; **cervids** (some small bovids may approach this form).
• The radius is rounded to oval in section, whilst the separate ulna is triangular in section; **pig, human, dog, cat.**

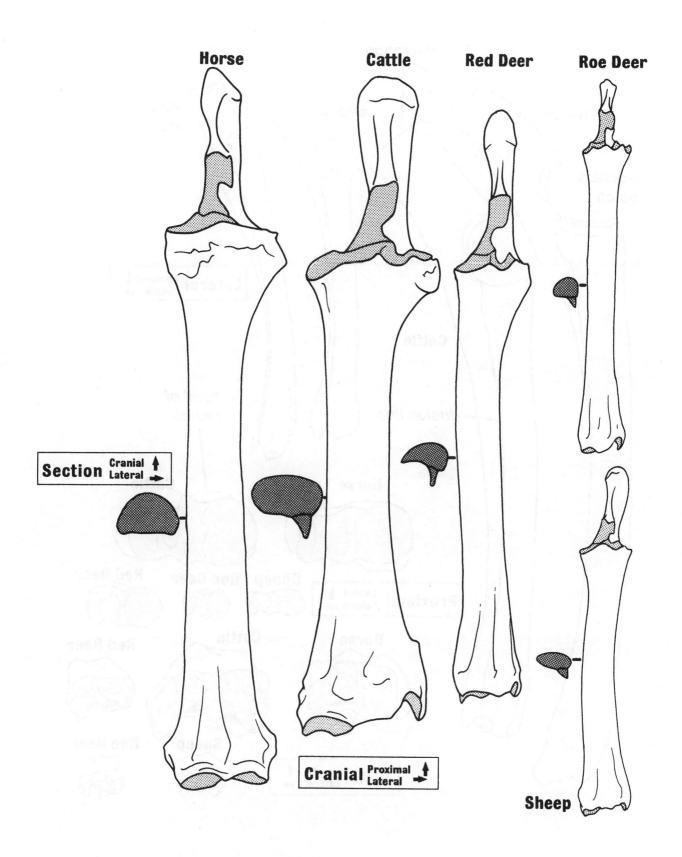

Horse **Cattle** **Red Deer** **Roe Deer**

Section Cranial ↑ Lateral →

Cranial Proximal ↑ Lateral →

Sheep

FIGURE 35. COMBINED RADIUS-ULNA (PART 1)

Cranial views, with mid-shaft section, for the left radius-ulna of horse, large bovid (cattle), large cervid (red deer), small cervid (roe deer), small bovid (sheep). Approximately half life-size.

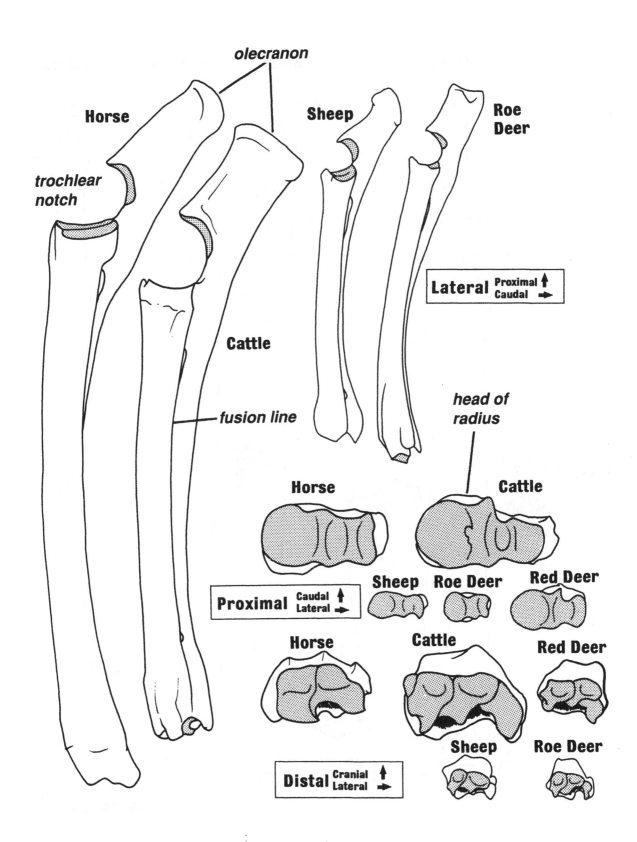

olecranon

Horse

trochlear notch

Sheep

Roe Deer

Cattle

Lateral Proximal ↑ Caudal →

fusion line

head of radius

Horse **Cattle**

Sheep **Roe Deer** **Red Deer**

Proximal Caudal ↑ Lateral →

Horse **Cattle** **Red Deer**

Sheep **Roe Deer**

Distal Cranial ↑ Lateral →

FIGURE 36. COMBINED RADIUS-ULNA (PART 2)
Lateral, proximal and distal views for the left radius-ulna of horse, large bovid (cattle), large cervid (red deer), small bovid (sheep) and small cervid (roe deer). Approximately half life-size.

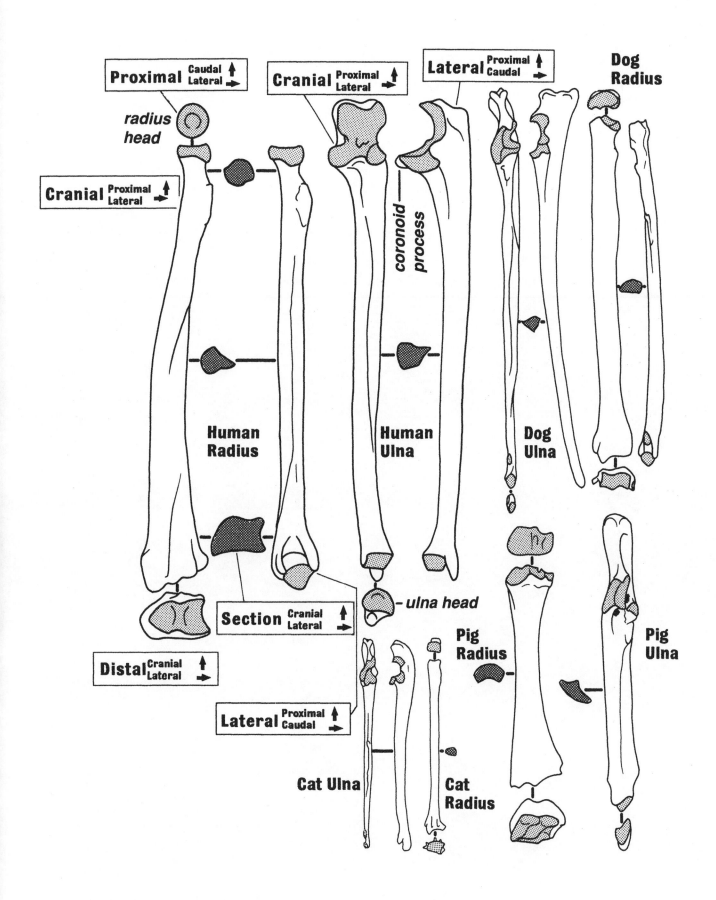

Proximal Caudal ↑ Lateral →

radius head

Cranial Proximal ↑ Lateral →

Cranial Proximal ↑ Lateral →

Lateral Proximal ↑ Caudal →

Dog Radius

coronoid process

Human Radius

Human Ulna

Dog Ulna

Section Cranial ↑ Lateral →

– *ulna head*

Distal Cranial ↑ Lateral →

Pig Radius

Pig Ulna

Lateral Proximal ↑ Caudal →

Cat Ulna

Cat Radius

FIGURE 37. SEPARATE RADIUS AND ULNA

Lateral, cranial, proximal and distal views, with mid-shaft sections, for left radius and ulna of human, dog, pig and cat. Approximately half life-size.

• The radius is bulging and rounded in section with a prominent lateral/caudal groove; the ulna is markedly asymmetrical in section, and both bones have large flattened and rough areas for the ligaments which bind them together; **pig**.

• The radius has a 'D'-shaped section, the ulna is slender, and the shafts of both bones bear prominent bulging roughened areas; **dog**, **cat** (the cat radius and ulna are minute in section).

• The radius has an elliptical section to distal, and a round section to proximal; both radius and ulna have a teardrop-shaped mid-shaft section, due to ridges for the interosseous ligament; **human**.

Shaft tuberosities

Humans, with a fully separated radius and ulna, have long, high ridge-like tuberosities to which the interosseous membrane holding the bones together is attached. The quadrupedal animals have much broader roughened areas on both bones, where ligaments bind them tightly together. On fusion, these *fusion areas* are still marked by a sharp line along much of their length.

• The fusion area is narrow and tapers to a fine trace over the proximal two thirds of the shaft; **horse**.

• The fusion area is broader and continues throughout the length of the shaft; **bovids, cervids**.

• Broad, flattish, roughened areas extend the full length of both bones; **pig**.

• The radius has a flattened area and a raised ridge, whilst the ulna has a ridge and a prominent bulging midshaft tuberosity; **dog, cat**.

• Both the radius and the ulna have prominent sharp crests along the length of their shafts; **human**.

Olecranon

• Long and massive, relatively parallel-sided or tapering in lateral view; **horse, bovids, cervids, pig**.

• Relatively short (proximal-distal), deep cranial-caudal, often with a proximal groove dividing the tuberosity; **dog, cat**.

• Very small, represented only by an area of roughening; **human**.

Proximal articulation – radius head

• Round and button-like with a continuous broad rim (for articulation with the ulna), the shaft is markedly waisted to distal of the head and there is a prominent lateral/caudal tubercle on the shaft a little further to distal; **human**.

• Oval and somewhat button-like with a broad rim on the medial/caudal side, the shaft is waisted to distal and bears a lateral/caudal tubercle; **dog, cat**; (the tubercle is more prominent in cat than in dog).

• Trough-like, with two ridges separating a larger dished area to medial and two smaller dished areas to lateral; **horse, bovids, cervids, pig**.

• The ridges are very sharply defined, and the head is narrower medial-lateral; **cervids**.

• The ridges are sharply defined, but the head is broader medial-lateral; **bovids**.

• The ridges are lower and smoother, and the head is broad medial-lateral; **horse**.

• The ridges are lower and smoother than in bovids, cervids and horse, and the head is narrower medial-lateral and deeper cranial-caudal, giving it a squarer proximal outline; **pig**.

Proximal articulation – trochlear notch of ulna

• A broad (medial-lateral) main articulation which rises to a prominent coronoid process, with a notch and cresentic facet to lateral for the head of the radius; **human**.

• A narrow (medial-lateral) main articulation which rises to a prominent coronoid process, with a prominent notch and crescentic lateral facet for the head of the radius; **dog, cat**.

• The **cat** coronoid process is markedly more prominent than that of **dog**, and has a characteristically distal tilted tip.

• A relatively narrow (medial-lateral) main articulation, with no real coronoid process and, on the cranial surface in unfused bones, twin oval facets for the radius; **horse, bovids, cervids, pig**.

• In cranial view, the notch articulation has a prominent re-entrant in the lateral side; **horse, bovids, cervids**.

Distal articulation

• Smooth and trough-like, with a lip-like facet for the head of the ulna to lateral and rising to a pointed process on the medial side; **human distal radius**.

• Button-like, with a raised process; **human ulna head**.

• A dished, oval distal facet, with a small facet to lateral for the head of the ulna; **dog, cat radius**.

• The distal end of the bone is blunt and rounded, and bears two rounded facets; **dog, cat ulna**.

• The distal ends of the radius and ulna together make up a complex of interconnected facets; two sinuous ridges cross the radius articulation obliquely, dividing it into two bulging medial areas and a lower lateral/caudal area, to which a dished facet on the head of ulna is added when the two bones are held together or fused; **bovids, cervids, pig**.

• The ridges on the distal articulation are very sharply developed; the area of the articulation as a whole is small relative to the size of the shaft; in mature animals the ulna is fused with the radius articulation to produce one continuous area; **cervids**.

• Similar sharply developed ridges, the area of articulation is relatively larger than in cervids; in mature animals the fused ulna again forms one continuous area of articulation with the radius; **bovids**.

• Less sharply defined ridges; the radius articulation is deeper cranial-caudal, giving it a squarer distal outline; the ulna articulation is separate; **pig**.

• The radius articulation is divided by a cranial-caudal ridge into two bulging areas; **horse**.

INNOMINATE BONE (FIGS. 38 & 39)

This is the most proximal bone of the hindlimb, bound with its neighbouring innominate bone to the sacrum of the spine, to make up the *pelvis*. Each innominate is centred on the *acetabulum*, a strongly buttressed socket for the spherical head of the femur. From the cranial part of this arises the portion of the innominate known as the *ilium*, narrowing into a *"neck"* and spreading out into the *iliac crest*. From the caudal edge of the acetabulum arises the *ischium*, which develops an *ischial tuberosity* of varying shapes on its caudal end. The third part of the innominate, the *pubis*, which runs medially from the acetabulum, is usually lost in archaeological specimens.

Acetabulum

The articular surface of the acetabulum describes an incomplete circle which surrounds a deep pit (the *acetabular fossa*), and is divided on its medial/caudal side by the *acetabular notch*. Just to cranial of the acetabulum are two rough depressions for the attachment of the rectus femoris muscle, called here *supra-acetabular fossae*. The more medial/cranial fossa is usually the more pronounced of the two.

• The acetabular notch is almost closed; **bovids, cervids, pig**.

• The acetabulum has a large diameter (>50mm); **large bovids, moose**.

• Intermediate acetabulum diameter (30-50mm); **red deer, caribou, fallow deer, pig**.

• Small acetabulum diameter (20-30mm); **roe deer, small bovids**.

• The rim of the acetabulum projects markedly and the acetabular fossa is usually deep; **pig**.

• The medial-cranial supra-acetabular fossa is variable, but it is usually more prominent in **bovids** and **cervids** than it is in **horse**.

• The acetabular notch is widely open and the acetabular fossa is narrow, giving a 'V'-shaped outline to the acetabulum in lateral view; **horse**.

• The acetabular notch is widely open and the acetabular fossa is relatively wide, giving the acetabulum as a whole a 'U'-shaped outline in lateral view; **human, dog, cat**.

• Intermediate to large (30-50mm+) acetabulum diameter; **human**.

• Small (15-30mm) acetabulum diameter; **dog**.

• Very small (around 10mm) acetabulum diameter; **cat**.

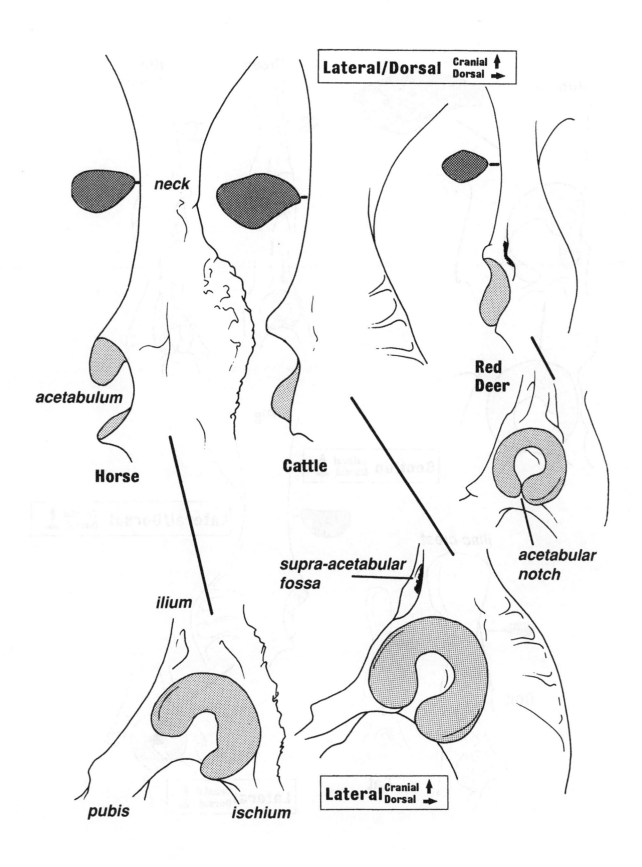

Lateral/Dorsal Cranial ↑ Dorsal →

neck

acetabulum

Horse

Cattle

Red Deer

supra-acetabular fossa

acetabular notch

ilium

pubis

ischium

Lateral Cranial ↑ Dorsal →

FIGURE 38. LARGE INNOMINATES
Lateral/dorsal views of acetabulum and ilium neck region, with lateral detail of acetabulum and a neck section, for left innominate of horse and large bovid (cattle). Approximately half life-size.

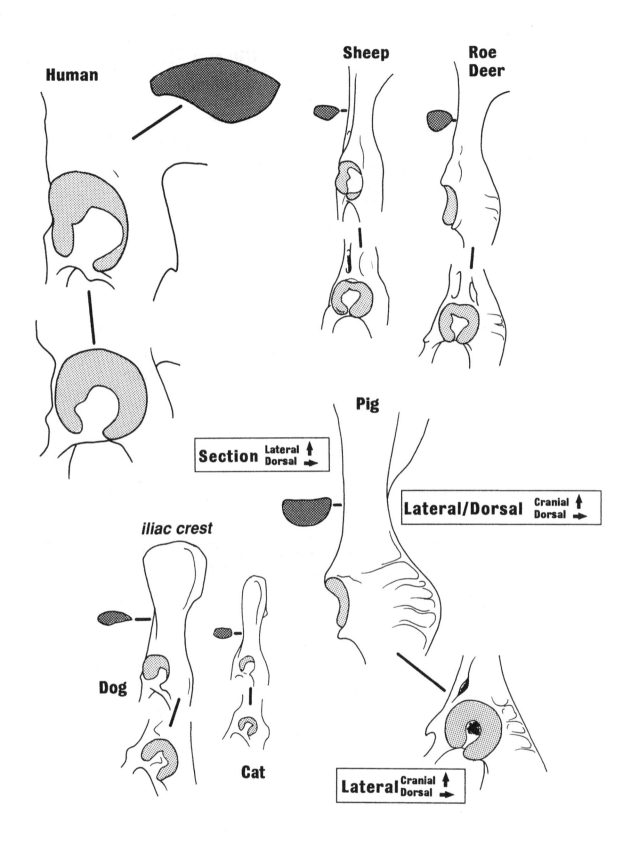

FIGURE 39. SMALL INNOMINATES

Lateral/dorsal views of acetabulum and ilium neck region, with lateral detail of acetabulum and a neck section, for left innominate of human, small cervid (roe deer), small bovid (sheep), dog, cat and pig. Approximately half life-size.

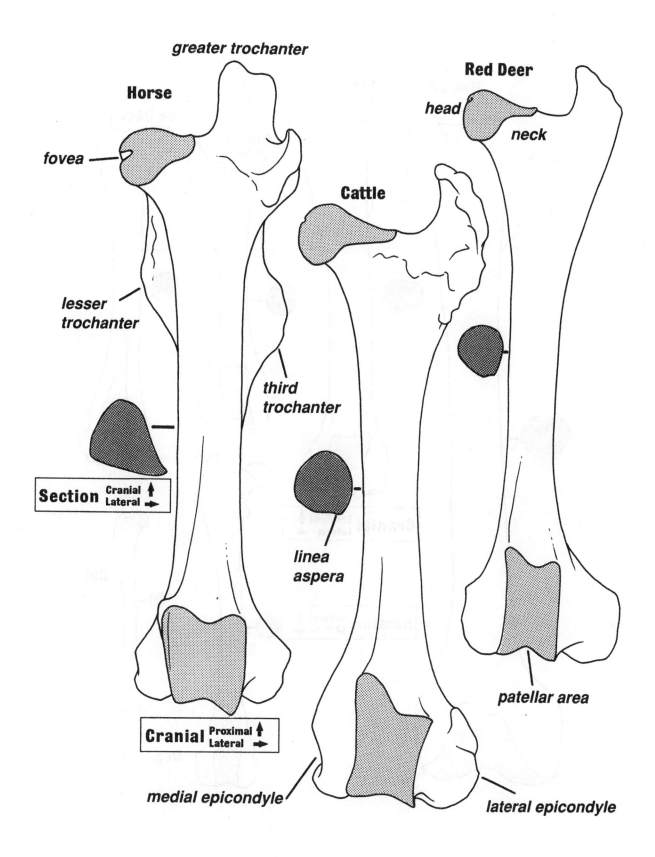

greater trochanter

Horse

fovea

lesser
trochanter

third
trochanter

Section Cranial ↑ Lateral →

Cranial Proximal ↑ Lateral →

medial epicondyle

Cattle

linea
aspera

Red Deer

head

neck

patellar area

lateral epicondyle

FIGURE 40. LARGE FEMUR

Cranial view, with mid-shaft section, for left femur of horse, large bovid (cattle) and large cervid (red deer). Approximately half life-size.

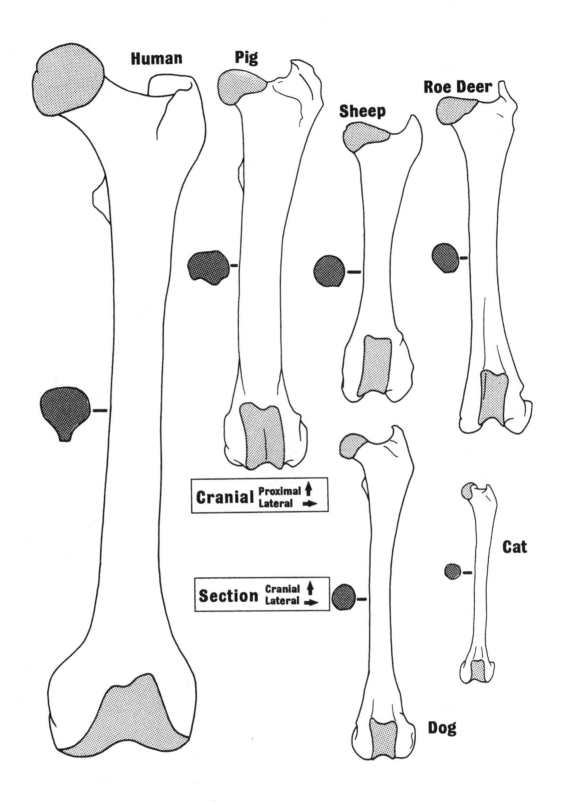

FIGURE 41. SMALL FEMUR
Cranial view, with mid-shaft section, for left femur of human, pig, small bovid (sheep), small cervid (roe deer), dog and cat. Approximately half life-size.

Neck of ilium

• The neck is broad dorsal-ventral, but very short caudal-cranial and surmounted by a massive, curved, spreading iliac crest; **human**.

• The neck is long caudal-cranial, quite broad dorsal-ventral (relative to acetabulum size), with a flattened section and surmounted by a narrow iliac crest; **dog, cat** (in the cat the crest is little broader than the neck).

• The neck is long caudal-cranial, slender dorsal-ventral (relative to acetabulum size), with an oval section and surmounted by flaring wings of iliac crest; **horse, bovids, cervids, pig**.

• In **cervids**, the neck is slightly more slender relative to its length than in **bovids**.

• In the **pig**, the neck bears a prominent ridge, giving it a 'D'-shaped section.

FEMUR (FIGS. 40-43)

The femur is the most proximal long bone of the hindlimb. In many mammals it has the most robust shaft of any long bone in the body. A prominent tuberosity, the *linea aspera*, runs down the caudal side of the shaft and divides as it reaches the distal end into two *supra-condyloid ridges*. At the proximal end of the shaft, the *femoral neck* arches out to medial and bears on its end the bulging *head* of the femur, which forms a 'ball-and-socket' joint with the acetabulum of the innominate bone. The proximal end of the shaft also bears tuberosities – the *greater trochanter* at the extreme proximal end, the *lesser trochanter* on the medial side just below the neck and (in horses), the *third trochanter* spreading out to lateral. Greater and lesser trochanters are connected, on the caudal side, by the *trochanteric crest* behind which there is often a deep *trochanteric fossa*. The distal end of the femur swells out to accommodate a broad articular area. On the cranial surface is the *patellar area*, which has a proximal-distal groove in which the patella (kneecap) runs. Attached to this and curving over the distal end of the bone, round to the caudal side, are

the *medial condyle* and *lateral condyle* which form joints with the condyles of the tibia. In between the condyles is the deep *intercondylar fossa* and on either side of them are moderately prominent tuberosities, the *medial epicondyle* and the *lateral epicondyle*.

Size and robustness

• Very long (proximal-distal) and robust; **large horse, large bovids**.

• As long, but more slender; **human**.

• Long and robust; **small horse, small cattle**.

• Long and somewhat more slender; **moose, red deer, caribou, fallow deer**.

• Short; **roe deer, small bovids, large dog**.

• Very short; **small dog, cat**; (the cat is usually around 100mm long).

• **Dog** and **cat** femur are quite similar in overall form (not size) to the **human** femur.

Shaft section and linea aspera

• The midshaft section outline is rounded, asymmetrical and with a marked angle formed by the distal tail of the third trochanter; the linea aspera is represented by a broad area of roughening; **horse**.

• The midshaft section is round and, in most specimens, shows a clear asymmetrically placed ridge at the linea aspera; **bovids, cervids, pig, dog, cat**.

• In **cervids** the section is smaller for a given length of shaft and the linea aspera more prominent than it is in **bovids**.

• The shaft is stout, with a broad and prominent linea aspera outlined by a groove on its medial side; **pig**.

• An asymmetrical section not unlike that of the bovids, but the linea aspera is less well developed; **dog, cat**.

• The section outline is symmetrical and the linea aspera is very prominent; **human**.

Head and neck

The head bears, on its medial side, a prominent pit called the *fovea*.

• The fovea is an isolated pit; **bovids, cervids, human, pig, dog, cat**.

• The fovea forms a deep 'V'-shaped indentation from the distal border of the head; **horse**.

• The head has a large diameter (*c.* 40mm), is extremely ball-like and bulges out from the neck on all sides; **human**.

• The head is similarly bulging and ball-like, but in some specimens it spreads with a very small margin onto the proximal surface of the neck; the head diameter is small (<30mm) and the neck is distinctly waisted below the head; **dog, cat**.

• The head is again bulging and ball-like, but it spreads just a little more onto the proximal border of the neck; the head diameter is small (*c.* 20mm) and the neck is not distinctly waisted; **pig**.

• The head spreads widely onto the proximal border of the neck, and there is little waisting of the neck; **horse, bovids, cervids**.

• The head spreads onto the neck slightly less in **horse** than in large **bovids**.

• **Cervids** have a slightly smaller head for a given shaft length than **bovids**.

Trochanters

• The large greater trochanter bears two swellings, the lesser trochanter is broad and flattened, a third trochanter is present as a broad protrusion, the trochanteric ridge is absent and the caudal area between the lesser and third trochanters is flat; **horse**.

• The large and tapering greater trochanter is connected by a prominent curved trochanteric ridge to a small but clearly outlined lesser trochanter; **bovids, cervids**.

• The greater trochanter is slightly less bulky (cranial-caudal) in **cervids** than in **bovids**.

• The greater trochanter is less bulky than in the bovids and cervids, and connected by a prominent curved trochanteric ridge to a small but prominent lesser trochanter; the more bulging head of the femur emphasises the dip of the neck in between; **pig**.

• The small greater trochanter is connected, by a curved and prominent trochanteric ridge, to a small but prominent lesser trochanter; **human, dog, cat**.

• The lesser trochanter is particularly prominent in **cat**.

Patellar area

• A low patellar area, with a broad and shallow groove running along a distal-proximal line; the medial and lateral borders spread out gradually to distal, with a broad connection to the condyles; the patellar area is relatively symmetrical but its lateral border bulges and is more prominent than the medial border; **human**.

• Still low, but the groove is deeper than in the human femur; the borders of the patellar area are higher and there is a narrower connection with the condyles; the area is relatively symmetrical, with the lateral border more prominent than the medial; **dog, cat**.

• The patellar area is high, with a deep groove and raised, swollen and parallel medial and lateral borders, with a narrow connection to the condyles (the lateral condyle is often separated); the area is obliquely arranged relative to the long axis of the femur and the medial border is more pronounced than the lateral; **bovids, cervids**.

• The borders are sharper and closer together in **cervids** than in **bovids**. The medial border is especially prominent in large bovids.

• The patellar area is similar to that of bovids, but it is broader, with less prominent borders and is less obliquely arranged relative to the long axis of the femur; **pig**.

• The patellar area is broad, widening to proximal, has a deep groove and swollen, prominent borders, the medial being only slightly more prominent than the lateral; **horse**.

• Two ridges arise from the patellar area borders and spread proximally onto the shaft; **horse, pig, human, dog, cat** (the last three show this feature less prominently).

• One ridge spreads proximally onto the shaft; **bovids, cervids**.

Condyles

• The condyles are smaller relative to the length of the femur in **cervids** than in **bovids**.

• The **dog** has prominent small oval facets to proximal of the condyles.

• The *lateral supracondylar fossa*, a rough depression to proximal of the lateral condyle, is especially deep and prominent in **horse**, prominent and broader in **bovids, cervids** and **pig**, and not prominent in **humans, dog**, or **cat**.

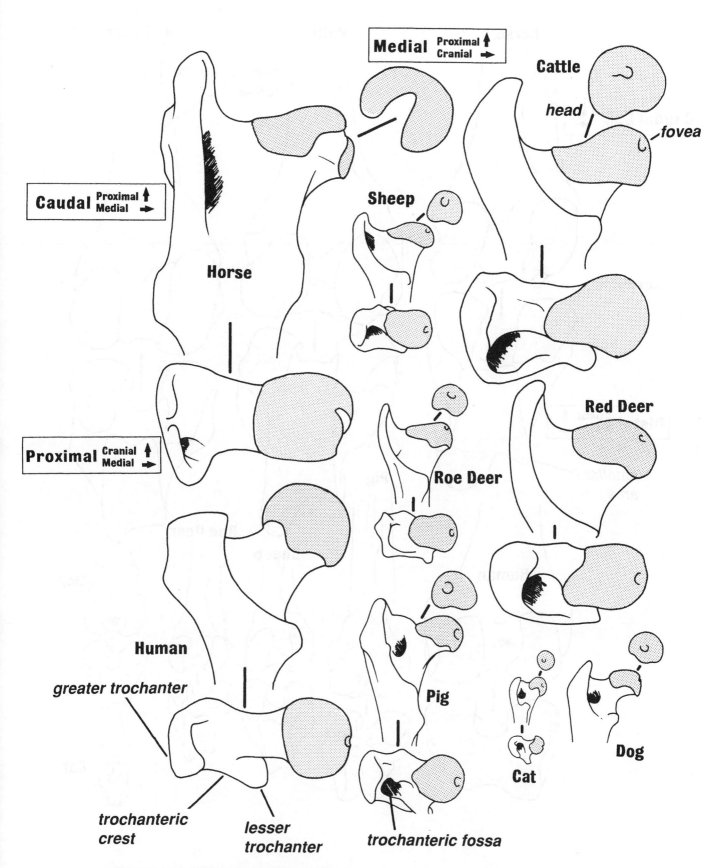

Medial Proximal ↑ Cranial →

Caudal Proximal ↑ Medial →

Proximal Cranial ↑ Medial →

Cattle

head

fovea

Sheep

Horse

Red Deer

Roe Deer

Human

greater trochanter

Pig

Cat

Dog

trochanteric crest

lesser trochanter

trochanteric fossa

FIGURE 42. FEMUR PROXIMAL ARTICULATION
Caudal view, with medial view of head, for left femur of horse, large bovid (cattle), human, small bovid (sheep), small cervid (roe deer), pig, large cervid (red deer), dog and cat. Approximately half life-size.

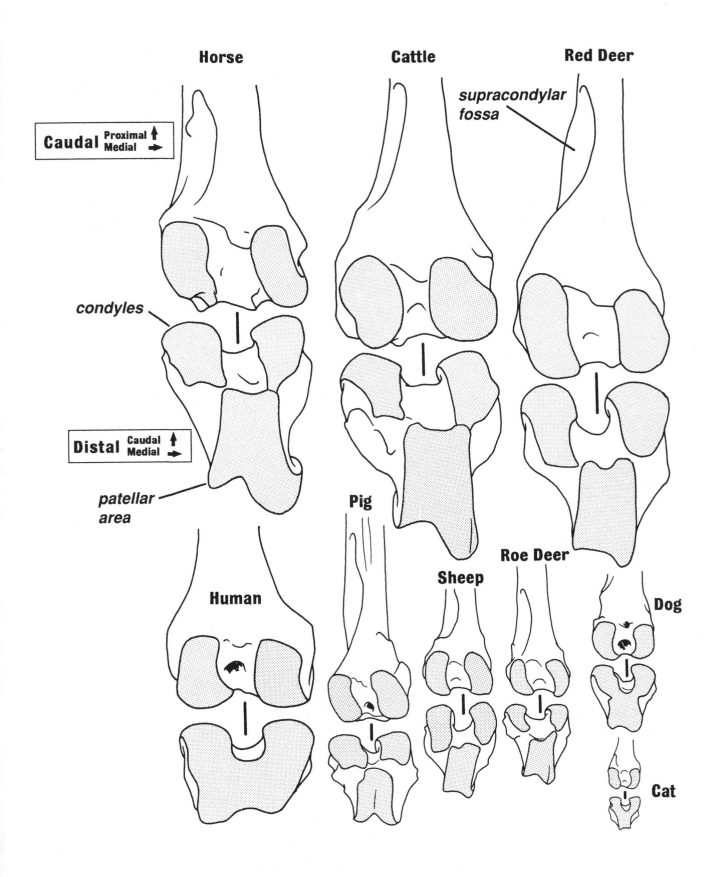

Horse   **Cattle**   **Red Deer**

Caudal | Proximal ↑ Medial →

supracondylar fossa

condyles

Distal | Caudal ↑ Medial →

patellar area

Pig

Roe Deer

Sheep

Human

Dog

Cat

FIGURE 43. FEMUR DISTAL ARTICULATION
Caudal and distal views of distal articulation, for left femur of horse, large bovid (cattle), large cervid (red deer), human, pig, small bovid (sheep), small cervid (roe deer), dog and cat. Approximately half life-size.

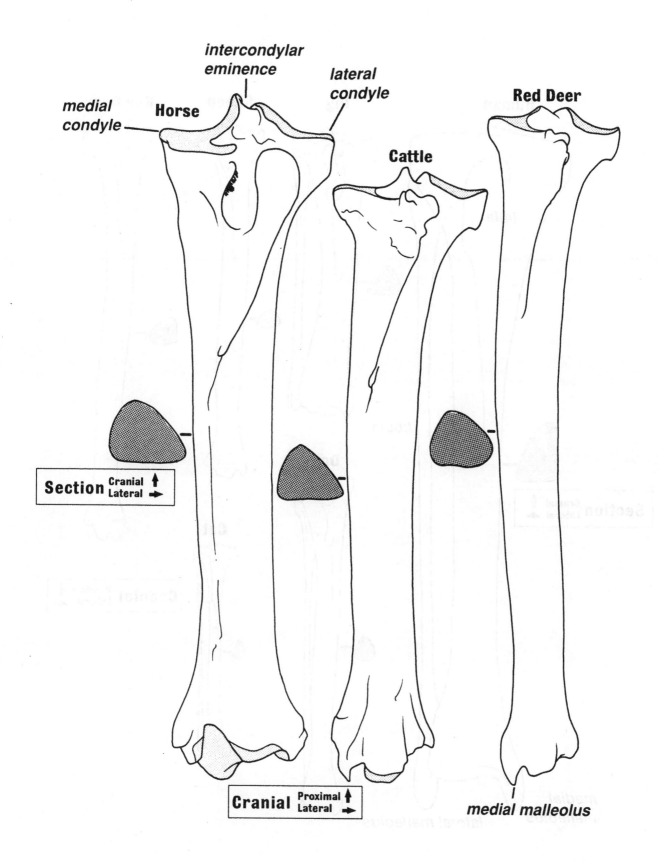

FIGURE 44. LARGE COMBINED TIBIA-FIBULA

Cranial views, with mid-shaft section, for left tibia-fibula of horse, large bovid (cattle) and large cervid (red deer). Approximately half life-size.

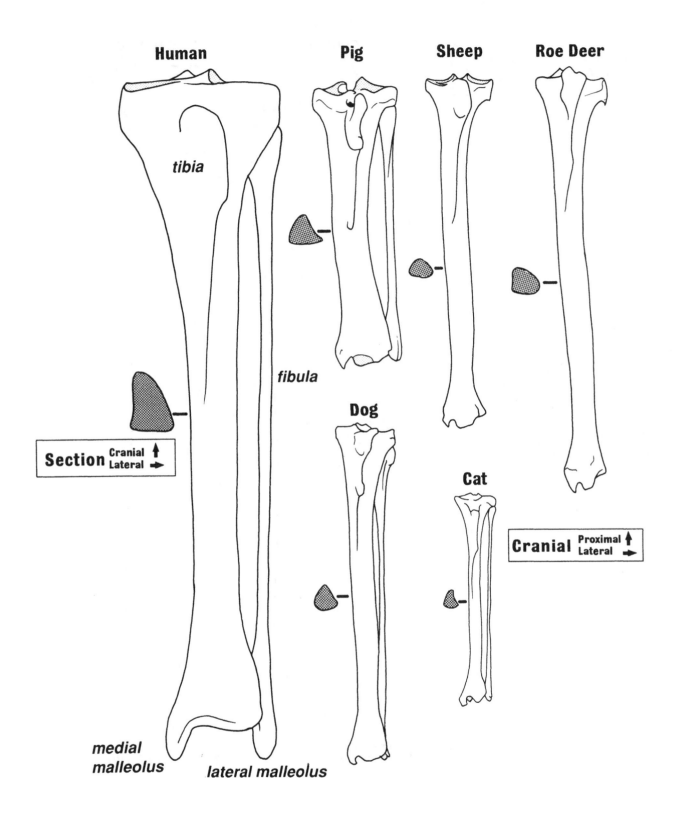

Human

Pig

Sheep

Roe Deer

tibia

fibula

Section Cranial ↑ Lateral →

Dog

Cat

Cranial Proximal ↑ Lateral →

medial malleolus

lateral malleolus

FIGURE 45. SMALL TIBIA AND FIBULA
Cranial views, with mid-shaft section, for left tibia and fibula of human, pig, dog and cat, left tibia-fibula of small bovid (sheep) and small cervid (roe deer). Approximately half life-size.

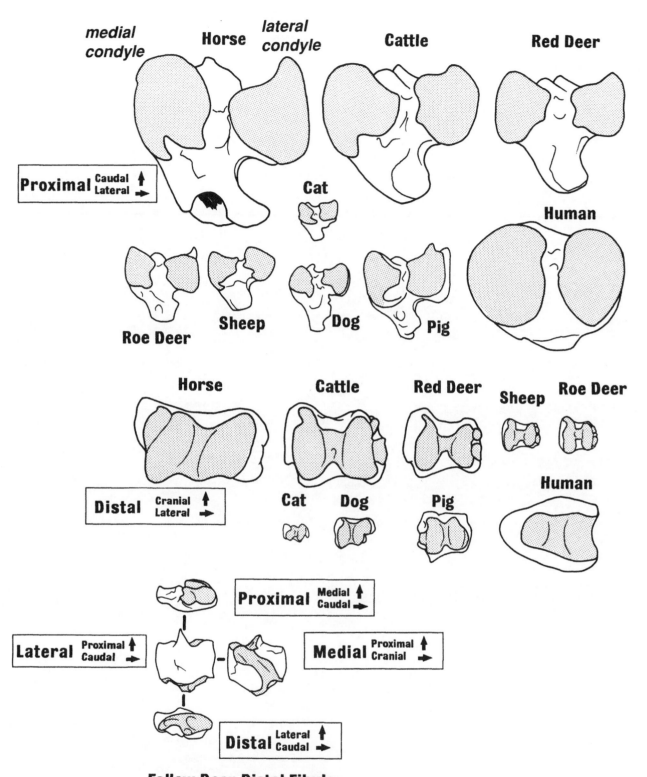

medial condyle · **Horse** · lateral condyle

Cattle

Red Deer

Proximal Caudal ↑ Lateral →

Cat

Human

Roe Deer

Sheep

Dog

Pig

Horse

Cattle

Red Deer

Sheep

Roe Deer

Distal Cranial ↑ Lateral →

Cat

Dog

Pig

Human

Proximal Medial ↑ Caudal →

Lateral Proximal ↑ Caudal →

Medial Proximal ↑ Cranial →

Distal Lateral ↑ Caudal →

Fallow Deer Distal Fibula

FIGURE 46. TIBIA PROXIMAL AND DISTAL ARTICULATION & DISTAL FIBULA IN BOVIDS/CERVIDS
Proximal and distal views for left tibia of horse, large bovid (cattle), large cervid (red deer), small cervid (roe deer), small bovid (sheep), dog, cat, pig and human.
The separate left distal fibula of an intermediate-sized cervid, fallow deer, has been selected as an example. Lateral, medial, proximal and distal views.
Approximately half life-size.

TIBIA–FIBULA (FIGS. 44-46)

The fibula is always much more slender than its massively built neighbour, the tibia. In some quadrupeds, the fibula is almost entirely absent and in others it is greatly reduced. The proximal end of the tibia is always the broadest part of the bone. On the relatively flat proximal surface, it bears two shallow, dished articular surfaces, the *medial condyle* and the *lateral condyle*, which curve up together in the centre of the proximal surface against the *intercondylar eminence*. Rising on the cranial side of the proximal end, a prominent tuberosity runs down the shaft as a crest, giving it a roughly triangular cross section. Where the fibula persists as a separate bone, a small facet for it is tucked into the overhang of the lateral/caudal corner of the proximal tibia. In some of the quadrupeds with a greatly reduced fibula, just a vestige is fused on at this point. The distal end of the tibia is swollen, bearing a trough-like articular surface for the joint with the talus. The border of this trough on the medial side bulges to distal as a small process, the *medial malleolus*. In those animals with a fully separate fibula, the end of this bone is also swollen into a *lateral malleolus* to pair the one on the tibia. Between them, they hold the talus (see below) into its joint.

Continuity of fibula
• The fibula is continuous throughout its length and remains unfused to the tibia throughout life; **pig, human, dog, cat**.
• The fibula is much reduced in its distal half, but remains a separate bone throughout the life of the animal; **horse**.
• The fibula is only present at its proximal extremity, a tiny vestige which fuses onto the tibia, and at its distal extremity, the lateral malleolus, which persists as a separate bone not unlike the tarsals; **bovids, cervids**.

Size and robustness of tibia
• Long and robust; **horse, medium-large bovids**..
• As long, but more slender; **human, moose, large red deer**.

• Intermediate length; **red deer, fallow deer, caribou, small cattle, large dog**.
• Short; **small bovids, dog**.
• Very short (overall length less than 150mm); **cat**.

Tibia midshaft section
• Asymmetrical section, with a rounded triangular outline; **horse, bovids, cervids, dog, cat**.
• The **horse** section is slightly less broad medial-lateral than that of the large **bovids**.
• The section in **cervids** is smaller for a given length of shaft than in **bovids**.
• The **dog** and **cat** have a relatively small section for the length of the tibia.
• Markedly triangular and strongly asymmetrical section; **human**.

Tibia tuberosity
• The tuberosity is relatively very prominent and angular, joining a high, thin and blade-like crest which extends down the proximal third to half of the cranial surface of the shaft, making the proximal outline into a narrow isosceles triangle; **dog, cat**.
• The tuberosity is massively developed and rounded, joining a crest which fades gradually into the shaft; **horse, bovids, cervids, pig**.
• The crest on the cranial surface of the shaft is more prominent and extensive in **cervids** than in **bovids**.
• There is a groove on the cranial tuberosity surface which produces two swollen areas; **horse**.
• The tuberosity is only moderately developed; **human**.

Condyles
• The condyle outlines are more rounded in the **human** tibia, and more irregular in the other animals.
• The intercondylar eminence is more prominent in **horse, bovids, cervids**.
• The area of the condyles is smaller relative to a given shaft length in **cervids** than in **bovids**.

Tibia distal articulation
• The articulation is divided by an oblique central ridge into two dished areas; **horse**.

CARPALS				
	Most medial			**Most lateral**
Proximal row	Radial (Scaphoid)	Intermediate (Lunate)	Ulnar (Triquetrum)	Accessory (Pisiform)
Distal Row	First (Trapezium)	Second (Trapezoid)	Third (Capitate)	Fourth (Hamate)

TARSALS				
	Most medial			**Most lateral**
Proximal row	Talus [Astragalus]		Calcaneus	
Centre	Central (Navicular)			
Distal row	First tarsal (1st Cuneiform)	Second tarsal (2nd Cuneiform)	Third tarsal (3rd Cuneiform)	Fourth tarsal (Cuboid)

• The articulation is divided by a cranial-caudally aligned central ridge into two dished areas; the medial malleolus is only moderately prominent; **bovids, cervids, pig, dog, cat.**

• There is an additional twin-lobed 'flange'-like area of articulation to lateral of the main articulation; **bovids, cervids.**

• The area of the articulation is smaller relative to the length of the shaft and its central ridge is more sharply defined in cervids than in bovids.

• The ridges and dished areas of articulation are more shallowly defined than in **cervids** and **bovids**, and the articulation is relatively broader cranial-caudal; **pig.**

• There is a pronounced step in the lateral border of the articulation; **dog, cat.**

• The distal articulation is a smoothly undulating facet, rising on its medial side to a prominent medial malleolus, and with a prominent groove on its lateral side for the fibula; **human.**

Fibula distal articulation

• No distal fibula; **horse.**

• The main articulation of the distal fibula consisting of a simple oval facet; **pig, human, dog, cat.**

• The distal fibula forms a small, separate, tarsal-like bone with multiple facets and a spine-like proximal process; **bovids, cervids.**

CARPALS (FIG. 47)

The maximum number of carpal bones is eight for each forelimb – four arranged in a proximal row and four in a distal row. The names given to them vary, and there are well established schemes for both human and veterinary anatomy. The names shown in the table are taken from Skerritt & Lelland (1984), with the equivalent in human anatomy given in brackets. Identification really requires experience beyond the level of this introductory book and details are not given here.

• **Human, pig, dog** and **cat.** All the bones are small and irregular and, although they are clearly distinguishable with experience, the distinctions are difficult to define precisely.

• **Bovid** and **cervid.** With the reduction to two metacarpals in these groups (see below), the distal carpal row is greatly modified. The Second and Third are fused together into one flat bone, and the Fourth is also markedly flattened. The Radial, Intermediate and Ulnar bones are compact and rectilinear in form.

• **Horse.** Radial, Intermediate, Ulnar, Second, Third and Fourth carpal bones are all rectilinear in form, fitting together into a compact block. The First is reduced to a small vestige.

TARSALS (FIGS. 48-50)

The maximum number of tarsal bones is seven – two bones in the proximal row, one central bone, and four bones in the distal row. Again the nomenclature is complex and Skerritt & Lelland

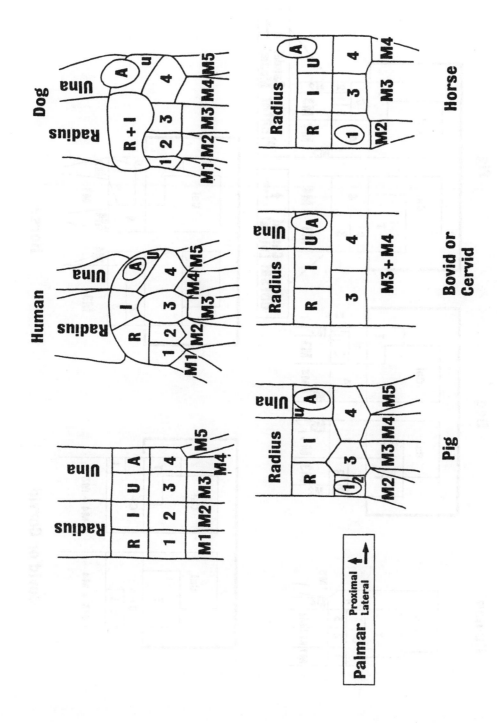

FIGURE 47. CARPALS

Block diagrams to show the arrangement of the right carpal bones, in dorsal view, for a generalised mammal, human, dog or cat, pig, bovids or cervids, and horse. Not to scale. Abbreviations for carpals: R - radial, I - intermediate, U - ulnar, A - accessory, 1 - first, 2 - second, 3 - third, 4 - fourth. Abbreviations for metacarpals: M1 - first, M2 - second, M3 - third, M4 - fourth, M5 - fifth.

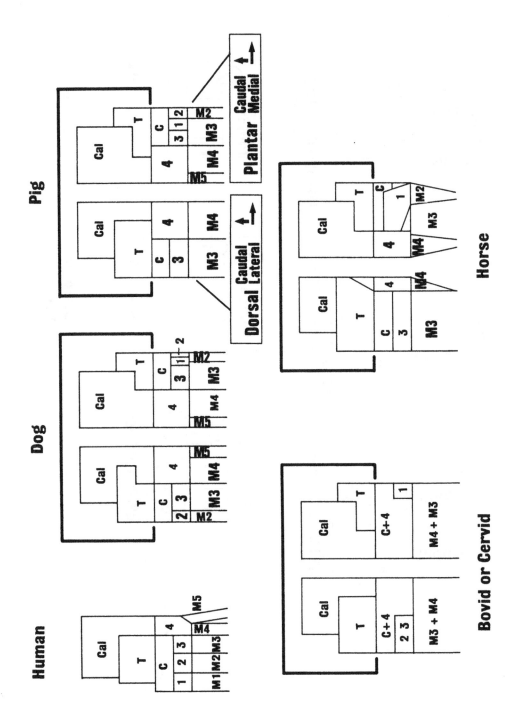

FIGURE 48. TARSALS

Block diagrams to show the arrangement of the left tarsal bones for human (in dorsal view only), and for dog, pig, bovid-cervid and horse (all in both dorsal and plantar view). Not to scale. Abbreviations for tarsals: Cal - calcaneus, T - talus, C - cuboid, 1 - first, 2 - second, 3 - third, 4 - fourth. Abbreviations for metatarsals: M1 - first, M2 - second, M3 - third, M4 - fourth.

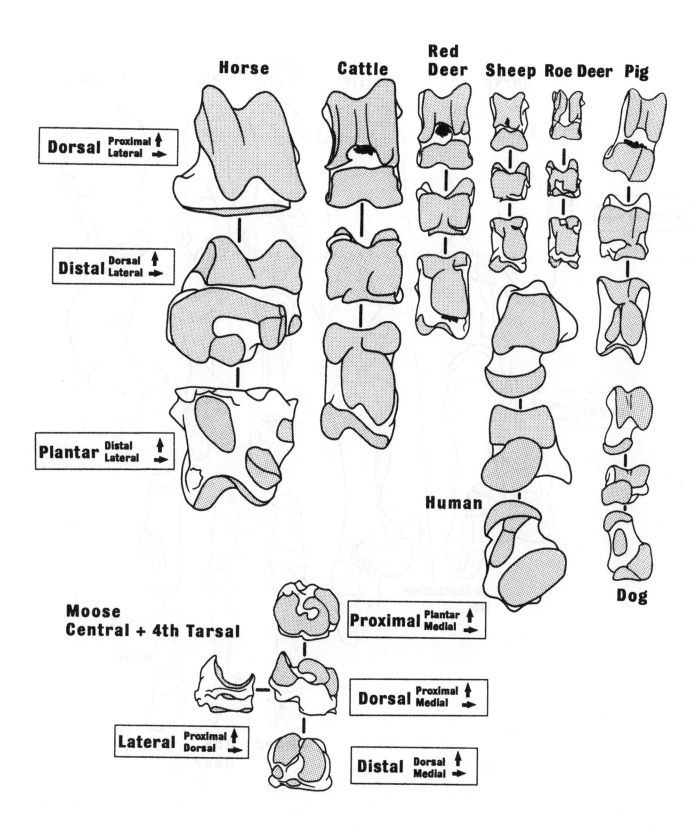

Horse **Cattle** **Red Deer** **Sheep** **Roe Deer** **Pig**

Dorsal $\frac{Proximal \uparrow}{Lateral \rightarrow}$

Distal $\frac{Dorsal \uparrow}{Lateral \rightarrow}$

Plantar $\frac{Distal \uparrow}{Lateral \rightarrow}$

Human

Dog

Moose
Central + 4th Tarsal

Proximal $\frac{Plantar \uparrow}{Medial \rightarrow}$

Dorsal $\frac{Proximal \uparrow}{Medial \rightarrow}$

Lateral $\frac{Proximal \uparrow}{Dorsal \rightarrow}$

Distal $\frac{Dorsal \uparrow}{Medial \rightarrow}$

FIGURE 49. TALUS & FUSED CENTRAL/FOURTH TARSAL

Dorsal, distal and plantar views for left talus of horse, large bovid (cattle), large cervid (moose), small bovid (sheep), small cervid (roe deer), pig, human and dog.

A large cervid (moose) has been selected as an example. Lateral, dorsal, proximal and distal views of left central-fourth tarsal.

Approximately half life-size.

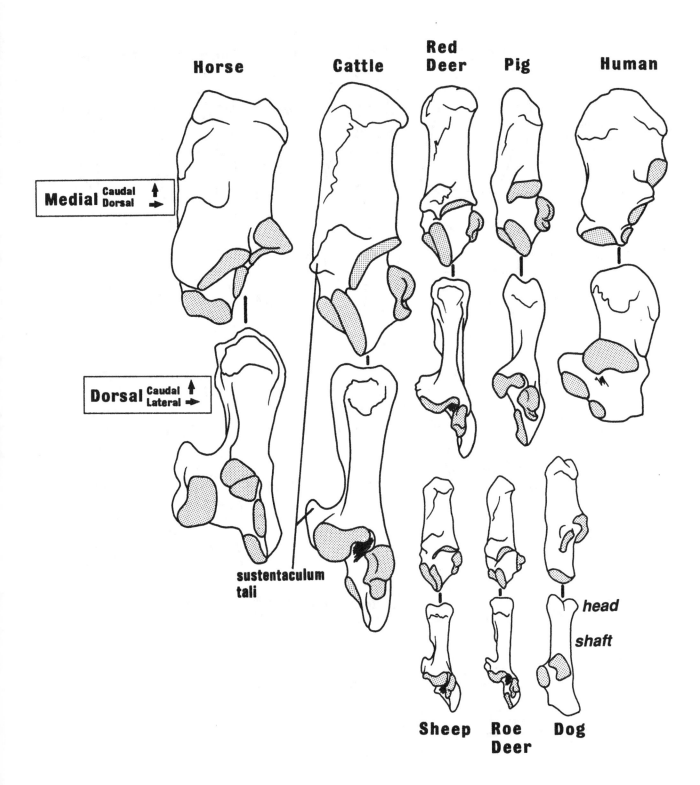

Horse   Cattle   Red Deer   Pig   Human

Medial Caudal ↑ Dorsal →

Dorsal Caudal ↑ Lateral →

sustentaculum tali

head

shaft

Sheep   Roe Deer   Dog

FIGURE 50. CALCANEUS
Medial and dorsal views for left calcaneus of horse, large bovid (cattle), large cervid (red deer), pig, human, small bovid (sheep), small cervid (roe deer) and dog. Approximately half life-size.

(1984) are followed, with common alternatives in square brackets, and alternative names from human anatomy in round brackets:

As with the carpals, although the bones are all distinctive, most distinctions are difficult to describe and are most easily studied in reference material. Instead, detail is given here only for the largest and most readily recognisable bones – the talus, the calcaneus and the fused central/fourth tarsal of bovids and cervids.

TALUS (FIG. 49)

This bone is compact in form, with two main articular facets. The large proximal articular surface for the tibia is invariably saddle or pulley-shaped, whereas the distal articulation for the central tarsal bone may be flat, domed or pulley-shaped. There are also several smaller facets on the plantar surface.

Size
- Large and robust; **horse, large bovids, moose, human.**
- Intermediate; **red deer, fallow deer, caribou, pig, small bovids, large dog.**
- Small to very small; **small dog, cat.**

Proximal articulation
- The ridges on the articulation are prominent and oblique relative to the long axis of bone; **horse.**
- The ridges are prominent, running parallel to the long axis of the bone; **bovids, cervids** (the ridges are particularly prominent in cervids).
- The ridges are organised in a similar way to bovids, but are less prominent; **pig.**
- The ridges are smoothly bulging, and still less prominent; **human, dog, cat.**

Distal articulation
- The facet is smooth and only slightly bulging; **horse.**
- The facet bulges into a swollen 'head'; **human, dog, cat.**
- There is a pronounced double ridge on the facet, mirroring the proximal articulation; **bovids, cervids.**
- Similar to bovids, but with the ridges less prominently developed and set obliquely to the long axis of the talus; **pig.**

Plantar articulation
- Two irregular facets; **horse, human, dog, cat.**

- One large, saddle-shaped facet, partly connected to the distal articulation; **bovids, cervids.**
- One large oval facet, split into two unequal parts; **pig.**

CALCANEUS (FIG. 50)

This bone has a swollen and roughened proximal end, the *calcaneus head*, to which the Achilles tendon is attached. The head is connected to the articular part of the bone by a *shaft* of varying length. The articulation consists of a complex of facets for the joints with the talus and central tarsal bones. In many mammals, there is a process on the medial side of the main articular facet for the talus called the *sustentaculum tali*.

Size
- Large and robust; **horse, large bovids, moose, human** (to some extent).
- Intermediate in size; **red deer, fallow deer, caribou, small bovids, pig, large dog.**
- Small to very small; **small dog, cat.**

Head and shaft
- The head is deep dorsal-plantar and the shaft is long proximal-distal; **horse.**
- The head is deep and the shaft is short; **human.**
- The head is not markedly deep, but the shaft is long; **bovids, cervids, pig, dog, cat.**
- The head is swollen dorsal-plantar relative to the shaft; **bovids, cervids.**

Distal articulation complex
- A prominent *sustentaculum tali* is present; **horse, bovids, cervids.**
- The lateral 'arm' of the articulation is long proximal-distal; **bovids, cervids, pig, dog, cat.**

CENTRAL AND DISTAL ROW OF HORSE, BOVIDS AND CERVIDS
- The central and third tarsals are flat and disc-like; **horse.**
- The second and third tarsals are fused together into a flat half disc-like bone; **bovids, cervids.**
- The central and fourth tarsals are fused together into a bone with a prominent step in its distal articulation (Fig. 49); **bovids, cervids.**

BOVID AND CERVID METAPODIALS (FIGS. 51-53)

In these families, the metacarpals and metatarsals are reduced to two in each limb, the third and fourth, and these are fused together into one structure from an early stage of development. The double origin of these composite third/fourth metapodials is, however, clear because a line is preserved down the length of the shaft near their point of fusion (particularly on the dorsal surface) and both distal and proximal joints are double in form. The general organisation of these metapodials is shown in Fig. 51, using cattle as an example. These bones distinguish themselves by their stout, composite shaft, double pulley-like distal articulation and almost flat proximal articulation. In young individuals whose distal epiphyses have not fused, the two elements of the distal articulation may be found in isolation, either separated or joined together.

DISTINGUISHING METATARSALS FROM METACARPALS

Shaft
• The **metatarsal** has a squarer outline in section; the **metacarpal** is markedly 'D'-shaped in section.
• The **metatarsal** has a markedly asymmetrical outline in section; the **metacarpal** is less so, but is still not completely symmetrical.
• The midline groove of the **metatarsal** is wider, deeper and more prominent than the midline groove of the **metacarpal**.
• The shaft of the **metatarsal** is relatively longer (proximal-distal) and more slender than the shaft of the **metacarpal** belonging to the same animal.

Proximal articulation
• The proximal surface of the **metacarpal** is an asymmetrical 'D' shape in outline.
• The proximal surface of the **metatarsal** is deeper dorsal-plantar and squarer in outline.
• The **metacarpal** has two facets, one larger than the other, between them covering almost the whole of the proximal surface.

• The **metatarsal** has two large triangular facets separated by a gap narrowing to dorsal, and two much smaller facets along the plantar border of the proximal surface.

Distal articulation
• For a **metacarpal** seen in lateral view, the point at which the most lateral element of the articulation meets the shaft is equally sharply indented and at a similar level on both dorsal and palmar sides.
• For a **metatarsal** seen in lateral view, the point at which the most lateral element of the articulation meets the shaft is less sharply indented and more proximally placed on the plantar side than it is on the dorsal side.

DISTINGUISHING LEFT FROM RIGHT
The third metapodial (more medial) is generally bulkier than the fourth metapodial and this gives rise to a number of distinctions between left and right.

Shaft
• The midline groove is placed slightly to lateral of the true midline.
• In a section, the bone to medial of the groove is bulkier and more protruberent than the bone to lateral. This is especially noticeable in metatarsals, with their more asymmetrical section outline.
• In metacarpals, the dorsal part of the section outline is curved convexly, and the palmar part is either flat or concave.
• In metatarsals, the dorsal part of the section is narrower (medial-lateral), more bulging and bears a more pronounced midline groove. By contrast, the whole of the plantar part is indented, sometimes very markedly so.

Proximal articulation
• In metacarpals, the dorsal part of the proximal outline is convex and the palmar part flat to very slightly concave. The 'D' shape of the outline is asymmetrical and is bulkiest on the medial side.

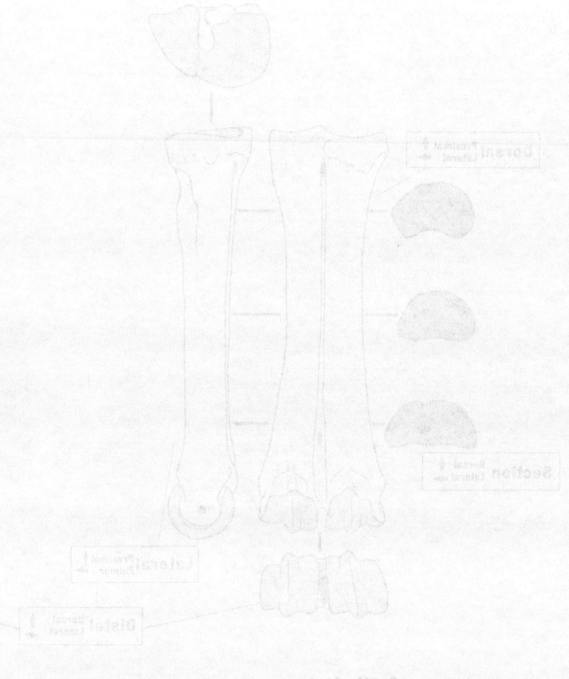

Cattle Metacarpal

FIGURE 61. LARGE BOVID METAPODIALS

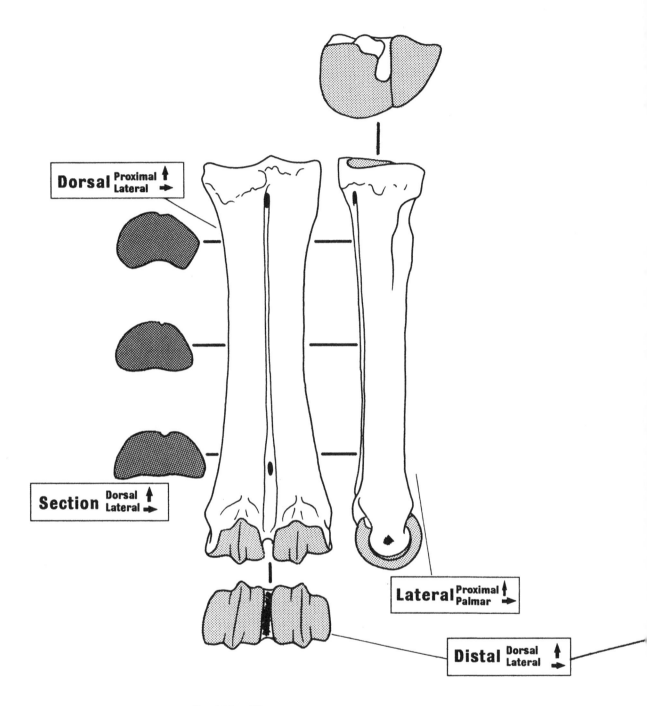

Dorsal Proximal ↑ Lateral →

Section Dorsal ↑ Lateral →

Lateral Proximal ↑ Palmar →

Distal Dorsal ↑ Lateral →

Cattle Metacarpal

FIGURE 51. LARGE BOVID METAPODIALS

Dorsal, lateral, proximal and distal views, with mid-, proximal and distal shaft sections, for left cattle metacarpals and metatarsals. Approximately half life-size.

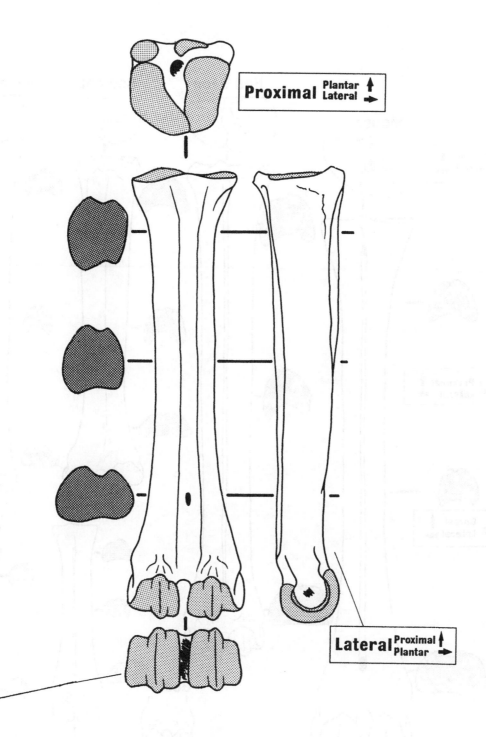

Metatarsal

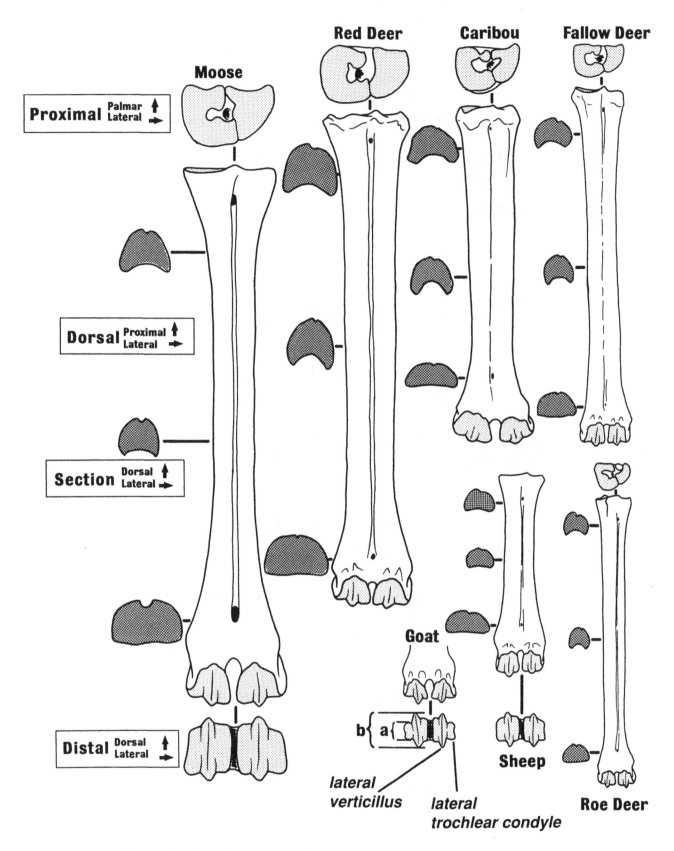

Moose

Red Deer

Caribou

Fallow Deer

Goat

Sheep

Roe Deer

Proximal Palmar ↑ Lateral →

Dorsal Proximal ↑ Lateral →

Section Dorsal ↑ Lateral →

Distal Dorsal ↑ Lateral →

b { a {

lateral verticillus

lateral trochlear condyle

FIGURE 52. CERVID AND BOVID METACARPALS
Dorsal, proximal and distal views, with mid-, proximal and distal shaft sections, for the left metacarpals of moose, red deer, caribou, fallow deer, roe deer and sheep. Detail of distal articulation in goat, with an illustration of the measurements described in the text. Approximately half life-size.

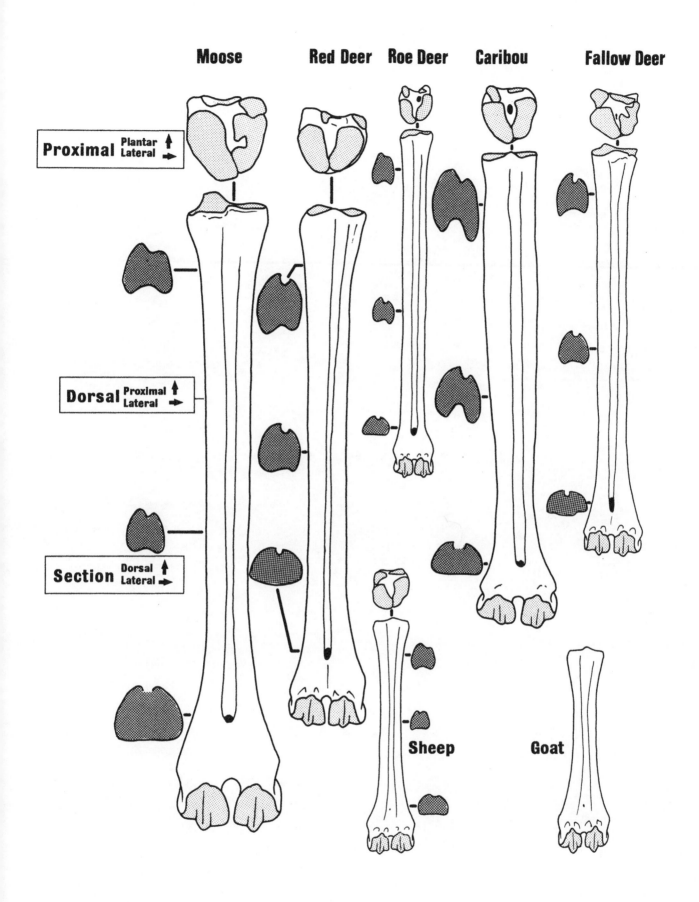

Moose **Red Deer** **Roe Deer** **Caribou** **Fallow Deer**

Proximal Plantar↑ Lateral→

Dorsal Proximal↑ Lateral→

Section Dorsal↑ Lateral→

Sheep **Goat**

FIGURE 53. CERVID AND BOVID METATARSALS

Dorsal and proximal views, with mid-, proximal and distal shaft sections, for the left metatarsals of moose, red deer, roe deer, caribou, fallow deer, sheep and goat. Approximately half life-size.

• In metacarpals, the medial articular facet is considerably larger than the lateral facet, so the dividing ridge between them is placed to lateral of the bone midline. The medial facet has a notch in it which may contain a pit.
• In metatarsals, the wedge-shaped proximal outline leans over to lateral, and the main medial articular facet is larger and extends slightly more to dorsal than the main lateral facet.

Distal articulation
• In both metacarpals and metatarsals, the most medial element of the articulation is bulkier than the most lateral element. This is clearest in distal outline and can usually be measured. The difference may not be so prominent amongst cervids as it is in bovids, but can be quite marked in the largest cervids such as moose and caribou.
• In cross section, the bone just to proximal of the distal articulation tends to be more flattened on its plantar or palmar surface than on its dorsal surface.

DISTINGUISHING BETWEEN TAXA
Bovid and cervid metapodials are amongst the most diagnostic of bones, so that both sheep and goat, and the different cervids can be described separately here.

Overall proportions
• The shaft is relatively stout for its length (proximal-distal); **bovids**.
• The shaft is relatively slender for its length (proximal-distal); **cervids**.
• Large and robust; **large bovids, moose, red deer**.
• Intermediate-sized; **fallow deer, caribou**, sometimes **small cattle**.
• Small; **small bovids, roe deer**.
• **Red deer, fallow deer** and **roe deer** are all rather similar in general proportions; their main difference is in their size.
• **Moose**, as well as being by far the largest of the **cervids**, has a pronounced expansion of its distal articulation relative to the shaft.

• **Caribou**, whilst retaining the main aspects of cervid form, diverges quite markedly from the others particularly in breadth (medial-lateral) and flattening (dorsal-palmar/plantar) of shaft and distal articulation.

Metacarpal shaft
• The mid-shaft section is only slightly concave, or even flat, in its palmar region; **bovids**.
• The mid-shaft section is markedly concave in its palmar region; **cervids**.
• The section of **caribou** is less deep (dorsal-palmar) relative to its width (medial-lateral), and more markedly concave (palmar) than other cervids.

Metatarsal shaft
• The midline dorsal groove is not sharply marked in a mid-shaft section, whilst the plantar surface is only slightly concave, or even flat; **bovids**.
• The midline dorsal groove is deep and sharply marked in mid-shaft section, with a deep concavity or indentation of the plantar surface; **cervids**.
• **Caribou** shows a particularly deep dorsal groove and plantar indentation, together with a very marked asymmetry of section.

Articulation
• The proximal and distal articulations are robust and broad; **bovids**.
• The proximal and distal articulations are less heavily built and narrower, relative to the shaft, than in bovids; **red deer, fallow deer, roe deer**.
• **Caribou** is distinguished by a compact proximal articulation with an expanded distal articulation bearing broad, rounded pulley-shaped articular surfaces.
• **Moose** is characterised by an expanded distal articulation.

Differences between sheep and goat
• In general, the metapodials are longer and more slender in **sheep** than in **goats**, although this is difficult to use for identification of fragmentary material.
• The twin ridges (*verticilli*) of the paired pulley-shaped distal articulation converge markedly to distal in the **goat**, but are more

METAPODIAL INDICES FOR SHEEP AND GOATS				
	Maximum for goat		Minimum for sheep	
	Boessneck	Clutton-Brock	Boessneck	Clutton-Brock
Metacarpal	63%	60.7%	63%	60.6%
Metatarsal	62.5%	62%	59%	62%

parallel in the **sheep**. The ridges are also more sharply defined in the **goat** than they are in **sheep**.

• The most lateral and medial elements (trochlear condyles) of the distal articulation are more notched-in relative to the verticilli of the articulation in **goat** than they are in **sheep**. Boessneck (1969) published a *metapodial index*, which uses two measurements taken with callipers (Fig. 52):

a. *Depth of medial trochlear condyle*. The maximum diameter between the dorsal and palmar/plantar surfaces of the medial trochlear condyle.

b. *Depth of medial verticillus*. This is the maximum diameter between the most prominent dorsal and palmar/plantar bulge of the medial verticillus.

To calculate the index, divide a by b and multiply the result by 100. It may be best to use such indices for classification of metapodials within each collection, rather than trying to define one universal rule. This is apparent when the results of Clutton-Brock *et al.* (1990) and Boessneck (1969) are compared (see table). Payne (1969) proposed different measurements that can be used in a similar way.

HORSE METAPODIALS (FIG. 54)

Horses balance on just the third digits of both manus and pes. The massive third metapodial is known as the *cannon bone* and it is flanked by the splinter-like remnants of the second and fourth metapodials, the *splint bones*.

DISTINGUISHING THIRD FROM SECOND AND FOURTH METAPODIALS

Shaft
• A large, robust, symmetrical shaft with a rounded section; **third**.

• A slender, tapering, asymmetrical, curved shaft with a triangular section; **second**, **fourth**.

Proximal articulation
• A large, relatively symmetrical proximal outline, bearing broad facets; **third**.
• Smaller, asymmetrical proximal outlines, bearing small facets; **second**, **fourth**.

Distal end
• A large condylar articular surface, with a prominent sagittal crest; **third**.
• A point, usually bearing a small nodule; **second**, **fourth**.

DISTINGUISHING THIRD METACARPALS FROM THIRD METATARSALS

Shaft
• **Metatarsals** are rounder in section, whilst **metacarpals** are more elliptical.
• **Metacarpals** have a broad, shallow groove on their palmar surfaces, whilst **metatarsals** have a narrow, prominent ridge, flanked by grooves.
• **Metatarsals** have a slightly longer, more slender shaft than **metacarpals**, but this difference is difficult to use for isolated specimens.

Proximal articulation
• The proximal outline is deeper dorsal-plantar in **metatarsals** than it is in **metacarpals**.
• The proximal end of a **metacarpal** is covered with facets, interrupted by grooves from medial and lateral.
• The proximal end of a **metatarsal** bears one large crescentic facet, enclosing a single rectangular, more plantar-placed facet.

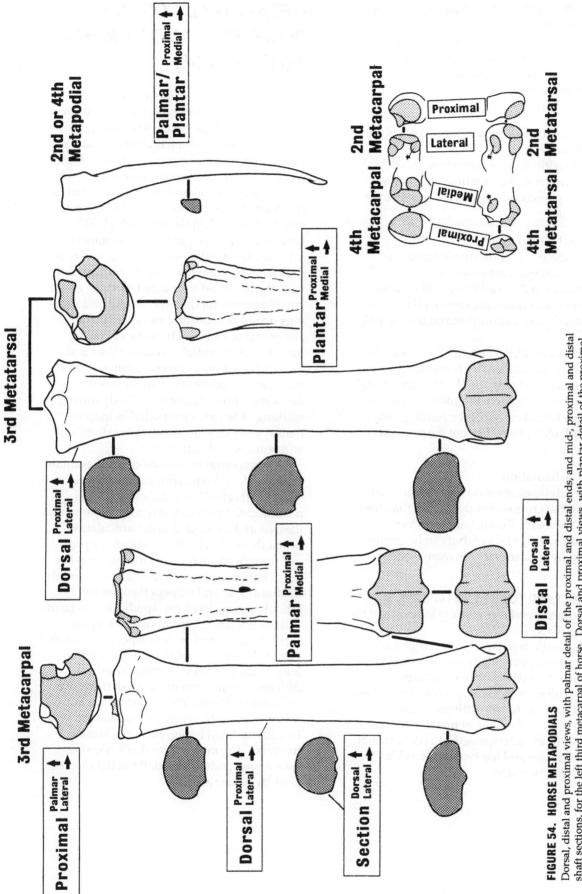

FIGURE 54. HORSE METAPODIALS

Dorsal, distal and proximal views, with palmar detail of the proximal and distal ends, and mid-, proximal and distal shaft sections, for the left third metacarpal of horse. Dorsal and proximal views, with plantar detail of the proximal end, and mid-, proximal and distal shaft sections, for the left third metatarsal of horse. Generalised palmar or plantar outline for a second or fourth metapodial of horse, with a mid-shaft section. Proximal and axial views of the proximal articulation for horse left second metacarpal, fourth metacarpal, second metatarsal and fourth metatarsal. The facets marked with "∗" are not always present. Approximately half life-size.

45

DISTINGUISHING LEFT FROM RIGHT THIRD METAPODIALS

The following features help to determine the dorsal, medial and lateral surfaces, which make it possible to establish which side of the body the specimen is from.

Shaft

• The most convex part of the section is the dorsal surface.
• The *nutrient foramen*, about one third the way down from the proximal end on the palmar/plantar side, is placed slightly over to medial.

Proximal articulation

• The tuberosity on the dorsal surface is slightly slewed over to medial.
• In metacarpals, in addition to the main proximal facet, there are two small facets on the lateral side, and only one on the medial side.
• In metatarsals, the ends of the crescentic main facet are marked by two smaller facets, the medial one being smaller than the lateral.
• In metatarsals, there is a tuberosity at the plantar/lateral corner of the rectangular, more plantar-placed facet of the proximal articulation.

Distal articulation

The condylar element on the medial side of the sagittal crest is slightly bulkier than the lateral element. This difference is very difficult to judge by eye, but can be demonstrated by measurement in many specimens.

SECOND AND FOURTH METAPODIALS

• The **metatarsals** are slightly longer than the **metacarpals**.
• The **fourth metatarsal** is the largest and most robust, particularly at its proximal end.
• The proximal outline of **metacarpals** is filled with joint facets – the **second** with two facets, and the **fourth** with one facet.
• The proximal outline of **metatarsals** has smaller facets interspersed by areas of tuberosity – the **second** has two facets, whilst the **fourth** has one facet.

METAPODIALS OF CATS, DOGS, PIGS AND HUMANS (FIGS. 55 & 56)

In these groups, more than one or two metapodials are retained in each manus or pes. **Cats** and **dogs** have five metacarpals in each manus and four metatarsals in each pes. The first metacarpal is, however, very much reduced in comparison with the others. The main axis of the pes and manus lies between the third and fourth metapodials, which are slightly longer than the second and fifth. Both manus and pes are almost symmetrically arranged either side of the central axis. **Pigs** have a somewhat similar arrangement, with the third and fourth metapodials much larger and stouter than the second and fifth – pigs do not possess first metapodials on either manus or pes. **Humans**, of course, have five metapodials in each manus and pes, with the first metapodial much stouter than the rest, particularly in the pes where it bears the whole weight of the body during walking. The first metapodial in human manus and pes is, in effect, the main axis of an asymmetrical structure.

A metapodial in any one of these animals is a small long bone, with a small to medium-sized shaft called the *body*, a flattish and multi-faceted proximal articulation known as the *base* and a rounded distal articulation called the *head*. The flat facets on the proximal surface of the base are for the carpal or tarsal bones, and there are small facets on the medial and lateral sides of the base for articulation between metapodials. The head usually has a ridge running down it, the *sagittal ridge*, which is developed to a variable extent. In theory, metatarsals are slightly longer and thinner than metacarpals, a difference which can distinguish them in a collection of bones from one skeleton but which is not sufficiently pronounced to identify isolated bones, except in human material. The most useful distinction between metapodials is usually in the form of joint facets on the bases.

46

LENGTH AND ROBUSTICITY
• Overall length (proximal-distal) is about the same for **humans**, **large dog** and **pig third** and **pig fourth**.
• A **small dog** may be half this length.
• A **large cat** may be as large as small dog, but most are smaller.
• The **third** and **fourth** metapodials of **pig** are markedly stout when compared with any human or large dog metapodials.
• The **second** and **fifth** metapodials of **pig** are only as long as a small dog, but more stoutly constructed; the fifth is slightly stouter than the second in any one pig.

BODY
• A long body, with a marked taper from an expanded base end (much roughened with points of attachment for ligaments) to a smaller diameter head end; waisted in the middle of the body, with a 'D'-shaped mid-body section; varying degrees of twist (passing from base to head); **human second, third, fourth** or **fifth metatarsal** (the twist gradually increases from a minimum in the second to a maximum in the fifth).
• A short body with a marked taper from a tiny base to a more bulbous head; the mid-body section is asymmetrical, 'D'-shaped in metatarsals and like one quadrant of a circle in metacarpals, with the broader of the two flattened section surfaces on the axial side; **pig second** or **fifth metapodial**.
• A long and stout body, with a less exaggerated taper from a smaller base to a bulbous head; the mid-body section is like one quadrant of a circle, having flattened surfaces to axial and palmar/plantar, and a curved surface to abaxial-dorsal (the axial surface is slightly concave and more roughened with ligament attachments than other surfaces); **pig third** or **fourth metapodial**.
• A long body, with marked waisting in the middle, an expanded base and a swollen head; prominent ridges define a long, flat triangular area on the dorsal surface of the body, with the point of the triangle towards the base (the flat dorsal surface is always recognisable in the otherwise irregular mid-body section); **human second, third, fourth** or **fifth metacarpal**.
• A long, stout body, with pronounced waisting, the base end is deep dorsal-plantar and the swollen head end is broader medial-lateral; the mid-body section is rounded 'D'-shaped; **human first metatarsal**.
• A moderate length body, often markedly waisted in the middle, with a broad, flat dorsal surface and a strongly curved palmar surface; the mid-body section has a flattened 'D'-shaped outline, with the flat surface to dorsal; **human first metacarpal**.
• A long to short body, slender, maintaining a similar width from the base to the head end, giving little appearance of waisting; straight or only slightly curved along its length; **dog, cat metapodial**.

BASE ARTICULAR FACETS
The base facets are the most useful part of the bone for identification, but do vary considerably between individuals of the same species. Some of the smaller facets are absent in some cases. Particular care is needed when defining facets in abraded archaeological specimens.

Human metatarsals
• **First** – bearing a large bean-shaped facet, with its concave side to lateral.
• **Second** – a triangular main base facet with two small tongue-like facets extending from the border of the main facet onto the lateral surface and one small facet extending onto the medial surface.
• **Third** – a triangular main facet, usually with a pronounced re-entrant on its lateral edge, with a large oval facet extending onto the lateral side and two small facets extending onto the medial side.
• **Fourth** – there is a marked groove and pit on the lateral surface of the body, just to distal of the base; the base bears a rectangular

47

main facet with a broad triangular extension onto the lateral side and a large tongue-like extension onto the medial side.
• **Fifth** – there is a marked tuberosity on the lateral side of the base; the base bears a rounded triangular main base facet, with a broad triangular extension onto the medial side.

Human metacarpals
• **First** – bearing a saddle-shaped facet (concavo-convex form); the broad sagittal bulge divides the facet into two halves, the medial half being the largest (but this difference is not marked).
• **Second** – the main facet is deeply folded along a dorsal-palmar axis; in proximal view it is roughly trapezoidal in outline, with a two-lobed extension onto the lateral side and a single small extension onto the medial side.
• **Third** – a broad, but prominent process (the *styloid process*) occupies the medial/dorsal corner; the main facet is triangular in proximal view and curves up onto the styloid process, and has two-lobed extensions onto both the medial and lateral sides.
• **Fourth** – the main facet has a roughly 'D'-shaped outline in proximal view, with a broad rectangular extension onto the lateral side and two small facets extending onto the medial side.
• **Fifth** – the main facet is concavo-convex and has a roughly trapezoidal outline in proximal view, with a broad triangular extension onto the medial side and a marked tubercle to lateral.

Dog and cat metatarsals
• The metatarsals have flatter main base facets than the metacarpals.
• **Second** – there is a marked tuberosity on the medial side of the body just to distal of the base; the main facet has a very narrow isosceles triangular outline in proximal view, but broadens slightly at its plantar apex with additional facets which extend round to enclose the plantar, medial and lateral surfaces of the apex; a small facet also extends onto the dorsal corner of the lateral surface.

• **Third** – the lateral side is deeply indented; the main facet is 'L'-shaped in proximal view, with one dorsal-plantar arm and one medial-lateral arm; the plantar tip of the dorsal-plantar arm is surrounded by small facets extending onto the medial, lateral and plantar sides; an additional small facet extends onto the dorsal corner of the medial side, whilst a larger oval facet is located in the dorsal roof of the indentation on the lateral side; **cat** is distinguished from **dog** by an indentation on the medial side of the main facet outline.
• **Fourth** – there is a prominent tuberosity on the medial side of the body just to distal of the base, with an oval facet on its dorsal surface; the main facet is roughly rectangular in proximal outline, with an indentation on its medial side; the plantar end of the facet is surrounded by small facets extending onto the medial, lateral and plantar sides; the lateral side bears an additional small facet and a tuberosity, under which is a pit whose roof is lined with yet another small facet.
• **Fifth** – the base has twin processes on its dorsal/lateral and plantar/lateral corners; there is a marked tuberosity on the medial side of the body just to distal of the base, with an oval facet on its dorsal surface; the main facet is irregular in outline, with facet extensions onto the sides of the two processes, and an additional oval facet on the plantar surface of the plantar/lateral process; **cat** is distinguished from **dog** by a great enlargement of the dorsal/lateral process.

Dog and cat metacarpals
• Metacarpals have more bulging main facets in medial and lateral view than metatarsals, usually with a broad sagittal groove running from dorsal to palmar.
• The lateral or medial (depending on which metacarpal) side is fringed continuously along its distal edge by a crescentic facet extension from the main base facet.
• **First** – small (very reduced in some individuals); there are tiny concavo-convex facets on both the base and the head.
• **Second** – bears a bulging, irregular base facet of roughly triangular outline in proximal view; a tuberosity bulges out beyond the

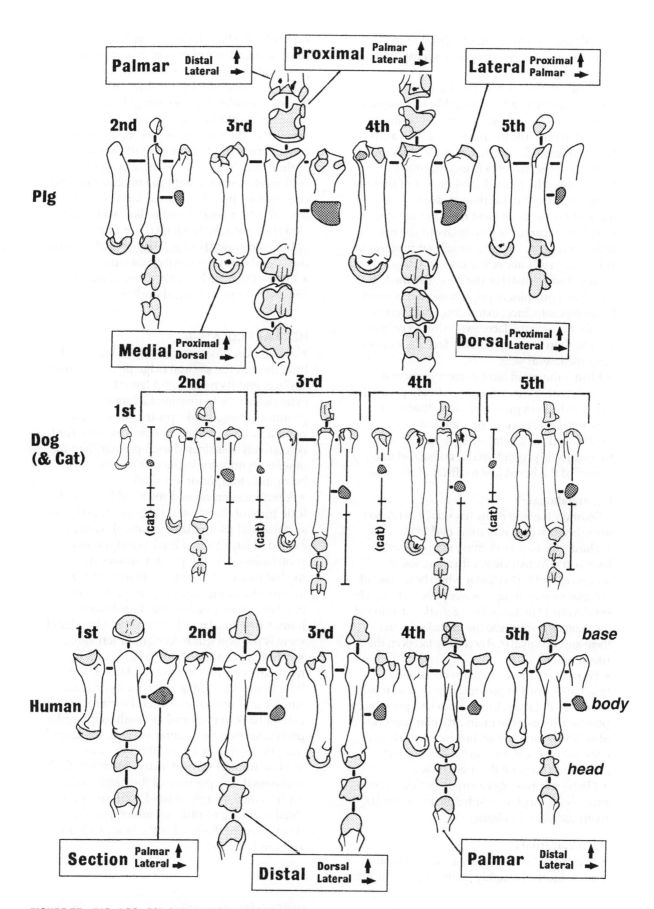

FIGURE 55. PIG, DOG, CAT AND HUMAN METACARPALS

Dorsal, medial, proximal and distal views, with lateral detail of the proximal articulation, palmar detail of the distal articulation and a mid-shaft section, for pig, dog and human left metacarpals. The dog illustrations incorporate scale bars which give a maximum and minimum length. Cat metacarpals are similar to those of dog, and they are only illustrated by scale bars to show their size range, with mid-shaft sections and, in the case of the fourth metacarpal, a dorsal detail of the proximal articulation which shows clear differences from dog. Approximately half life-size.

dorsal/medial corner just to distal of the main facet; there is a simple crescentic extension onto the lateral side.

• **Third** – the main base facet is narrow medial-lateral and has a roughly rectangular proximal outline; there is a crescentic facet extension onto the medial side and an additional tongue-like extension in the dorsal/medial corner; two facets extend onto the lateral side (the more dorsal-placed of these is relatively larger in **cat** than in **dog**).

• **Fourth** – the main facet has (on the right fourth metacarpal) a proximal outline not unlike a map of Africa; a small but prominent tuberosity rises above the dorsal/lateral corner, just to distal of the base; the medial side has a prominent pit, whose roof is lined by a crescentic facet extension, and there is another crescentic facet extension to palmar of this; there are two narrow facet extensions onto the lateral side.

• **Fifth** – the main facet is narrow medial-lateral, with a curved, but relatively parallel-sided outline in proximal view; there is a crescentic facet extension onto both the medial and the lateral sides; a long, narrow tuberosity is present on the dorsal/lateral corner of the body, just to distal of the base.

Pig metatarsals

• **Second** – the base is tiny (especially in relation to the head), with a very small oval facet.

• **Third** – the base is relatively narrow (medial-lateral) in proximal view, with a prominent process extending to plantar which bears a small articular facet on its tip; the main facet is relatively small, with a fan-like outline; a single extension of the main facet runs onto the medial side, and there are two small oval accessory facets on the lateral side.

• **Fourth** – the base is relatively broader (medial-lateral) in proximal view than it is in the third metatarsal, and has a less prominent plantar process; the main articular facet is also fan-like in outline, but more markedly concavo-convex; there are two irregular accessory facets on the medial side.

• **Fifth** – the base has a small, but distinctive process on its plantar side; the facet is relatively large and bulging.

Pig metacarpals

• **Second** – the base bears a small, concavo-convex facet of roughly bean-shaped outline.

• **Third** – the base bears a broad, concavo-convex main facet of irregular outline in proximal view; there are two small facet extensions onto the lateral side and one onto the medial side; there are irregular facet extensions onto the palmar side.

• **Fourth** – the base carries a broad, mostly convex main facet of roughly triangular outline in proximal view; there is one large, oval facet extension onto the medial side and one smaller, usually isolated facet on the medial side; a small tongue-like facet extension runs onto the medial corner of the palmar side and there is a larger oval facet on the lateral corner of the palmar side.

• **Fifth** – the base has only a small, mostly convex facet with an oval outline.

HEAD ARTICULATION

• A bulbous, broad and rounded head with little trace of the sagittal ridge on its articular surface, and its medial and lateral corners extending back to proximal as small but prominent tongue-like protrusions on the palmar surface; **human metacarpal** (the **first** is relatively less deep dorsal-palmar than the others; asymmetry in the distal view is the best guide to right or left side).

• A bulbous, broad and rounded head with little trace of the sagittal ridge on its articular surface, and its medial and lateral corners extended back to proximal as tongue-like protrusions onto the plantar surface, the medial tongue being larger than the lateral; the articular surface is strongly marked on its distal and plantar surface with broad grooves; **human first metatarsal** (asymmetry in the distal view is the best guide to right or left side).

• A bulbous and rounded head which is markedly compressed medial-lateral, with little trace of a sagittal ridge on its articular surface and medial and lateral corners which extend back to proximal as small tongue-like protrusions on the plantar surface, the lateral 'tongue' normally being larger than the medial; **human second**, **third**, **fourth** or **fifth** **metatarsal** (the position of the head on the body is increasingly rotated from second to third to fourth to fifth; asymmetry in the distal view of the head is the best guide to right or left side).

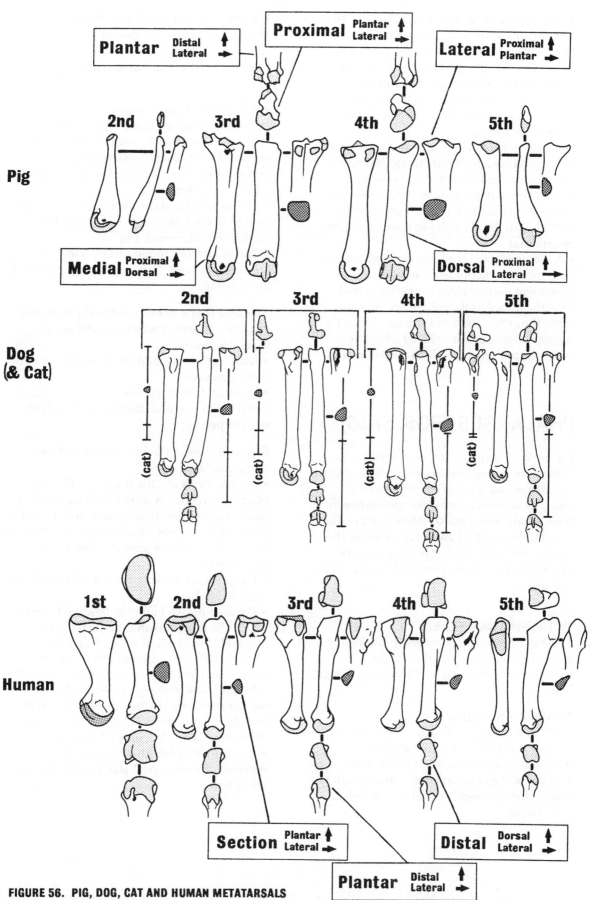

FIGURE 56. PIG, DOG, CAT AND HUMAN METATARSALS

Dorsal, medial , proximal and distal views, with lateral detail of the proximal articulation, plantar detail of the distal articulation and a mid-shaft section, for pig, dog and human left metatarsals. The dog illustrations incorporate scale bars which give a maximum and minimum length. Cat metatarsals are similar to those of dog, and they are only illustrated by scale bars to show their size range, with mid-shaft sections and, in the case of the third and fifth metatarsals, details of the proximal articulation which shows clear differences from dog. Approximately half life-size.

• A relatively small and narrow head with a prominent sagittal ridge, which fades out on the dorsal surface but protrudes to proximal on the palmar/plantar surface; **dog** or **cat metapodial** (a slight asymmetry in distal view, greatest in second and fifth metapodials, is a guide to the axial and abaxial sides of the head).
• A large and broad head, of marked asymmetry, and with a prominent sagittal ridge extending round from the dorsal to the palmar/plantar surfaces; in distal view, the axial articulation is deeper dorsal-palmar/plantar than the abaxial; **pig third** or **fourth metapodial**.
• A medium sized but bulbous head which is even more markedly asymmetrical, with a prominent sagittal ridge extending round from the dorsal to the palmar/plantar surfaces; the articulation on the axial side of the sagittal ridge is much reduced (effectively missing); **pig second** or **fifth metapodial**.

PHALANGES (FIGS. 57-59)

For most of the following descriptions, no distinction is made between the phalanges; of the manus and those of the pes. In most cases, there is no useful difference between them. Only in human skeletons is it possible to tell them apart reliably. Each digit of the manus and pes is made up of three (sometimes two) phalanges: proximal, intermediate and distal.

DISTAL PHALANX (FIG. 57)
This bone always has a strongly modified form, depending on whether the limb is terminated by nails, claws, hooves, or a single hoof.

Proximal articulation
• All have a single articular surface divided by a low ridge into two concavities.
• A ridge is symmetrically placed on the midline of the large articular surface, with concavities of roughly equal size on either side; **horse**.

• A ridge is asymmetrically placed (towards the axial side) on the articular surface, with concavities of unequal size either side; **bovids, cervids, pig, human, dog, cat**.

Distal form
• A prominent and symmetrical tuberosity fans-out around the distal end; **human**.
• The distal part of the phalanx consists of a curved, finger-like process, with a sharp groove at its base encircled by an upstanding frill of bone; **dog, cat**.
• The distal part of the phalanx consists of a wedge-shaped process with a triangular section; **bovid, cervid, pig**.
• The phalanx is dominated by a massive crescentic process, surrounding the proximal facet; **horse**.

Size and shape of human distal phalanges
• Large and well developed; **first** digit of **manus** and **pes**.
• Intermediate; **second, third, fourth** or **fifth** digit of **manus**.
• Small and reduced especially in proximal-distal length; **second, third, fourth** or **fifth** digit of **pes**.

Size and shape of bovid, cervid and pig distal phalanges
• For most **cervids**, the **third-fourth** digit phalanx is relatively long proximal-distal and narrow axial-abaxial, so **moose, red deer** and **fallow deer** are often the longest even though they may be smaller animals overall than large **bovids**.
• **Large bovid** phalanges are markedly broad axial-abaxial.
• **Caribou** have a highly individual form in which the **third-fourth** digit phalanx is relatively broad (axial-abaxial) and low (dorsal-palmar or plantar) when seen in section, and strongly curved along its length.
• **Small bovid** and **pig third-fourth** digit phalanx are roughly the same length proximal-distal, and are all considerably smaller than in the larger **bovids** and **cervids**.
• The **pig third-fourth** digit phalanx is markedly broader axial-abaxial than that of a small bovid.

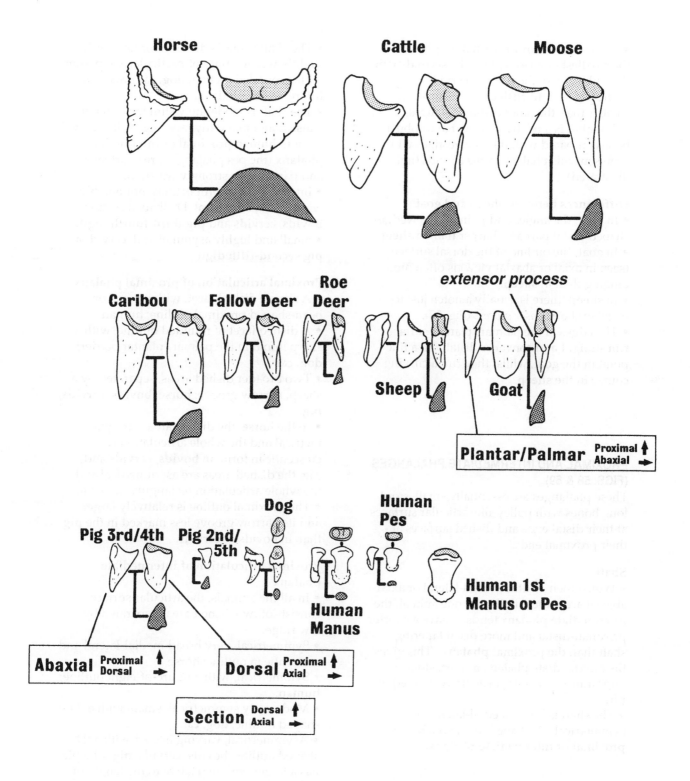

FIGURE 57. DISTAL PHALANGES

The distal phalanges of most mammals are difficult to distinguish between manus and pes, left and right. This figure therefore illustrates just generalised distal phalanges, except in the case of humans and pigs, where differences are readily apparent. Dorsal and abaxial views, with a mid-shaft section, for horse, cattle, moose, reindeer, fallow deer, roe deer, sheep, goat, pig third or fourth digit, and dog distal phalanges. The sheep and goat distal phalanges also have palmar (or plantar) views. Dorsal view and section of pig second or fifth distal phalanx. Medial, dorsal and proximal views, with a mid-shaft section for human left manus and pes phalanges (generalised). Dorsal view of a human left pes first digit phalanx. Approximately half life-size.

- The **roe deer third-fourth** digit phalanx is the smallest of all except for the **second-fifth** digit phalanx of the pig and cervids.
- The **pig second-fifth** digit phalanx is much smaller than the other distal phalanges, although it is a relatively broad and robust bone compared with the tiny vestiges which represent this phalanx in the cervids (not illustrated).

Differences between sheep and goat
- In **goat**, the angles and point of the wedge-shaped distal part are sharper than in **sheep**.
- In **goat**, the outline of the dorsal surface (seen in axial or abaxial view) is often more bulging than in **sheep**.
- In **sheep**, there is usually a notch just to distal of the *extensor process* (Fig. 57).
- The edges of the palmar/plantar surface run straight and converge distally to a fine point in the **goat**, whilst they run a curving course in the **sheep**.

PROXIMAL AND INTERMEDIATE PHALANGES (FIGS. 58 & 59)
These phalanges are essentially small, stout long bones with pulley-like articular surfaces at their distal ends and dished surfaces at their proximal ends.

Shaft
- Whilst their widths medial-lateral or axial-abaxial are rather similar for one animal, the **intermediate phalanx** tends to have a shorter proximal-distal and more distal tapering shaft than the **proximal phalanx**. This gives the intermediate phalanx a short, stout appearance which is particularly marked in **pig**.
- The shaft is broad medial-lateral and symmetrical 'D'-shaped in section; **horse proximal** or **intermediate phalanx**.

- The shaft is markedly slim for its length, slightly waisted but not particularly tapering, with a rounded section; **dog, cat proximal** or **intermediate phalanx**.
- The shaft is moderately slim for its length, waisted and markedly tapering to distal, oval section; **human proximal** or **intermediate phalanx** (the pes phalanges are shortened and particularly strongly waisted).
- Intermediate in proportions, not notably waisted, asymmetrical 'D'-shaped section; **bovids, cervids** and **pig third-fourth digit**.
- Small and highly asymmetrical; **cervid** or **pig second-fifth digit**.

Proximal articulation of proximal phalanx
- A simple dished facet, with an oval or bean-shaped proximal outline; **human**.
- A dished facet of rounded outline, with a sharp infold in the palmar/plantar border; **dog, cat**.
- Two broader dished areas, separated by a sharp, narrow groove; **horse, bovids, cervids, pig**.
- In the **horse**, the dished areas are symmetrical and the whole articulation is crescentic in form; in **bovids, cervids** and **pig**, the dished areas are asymmetrical and the whole articulation rectangular in form.
- The proximal outline is relatively larger and its narrow groove less marked in the **pig** than in bovids or cervids.

Proximal articulation of intermediate phalanx
- In all the animals, the articular surface consists of two dished areas separated by a low ridge.
- Symmetrical, very broad medial-lateral and crescentic in outline; **horse**.
- Symmetrical, with a figure-of-eight outline; **human**.
- Moderately symmetrical, small, with a 'D'-shaped outline; **dog, cat**.
- Asymmetrical, varying in size, with a 'D'-shaped outline; **bovids, cervids, pig**; (the pig has a larger outline relative to the length of the bone).

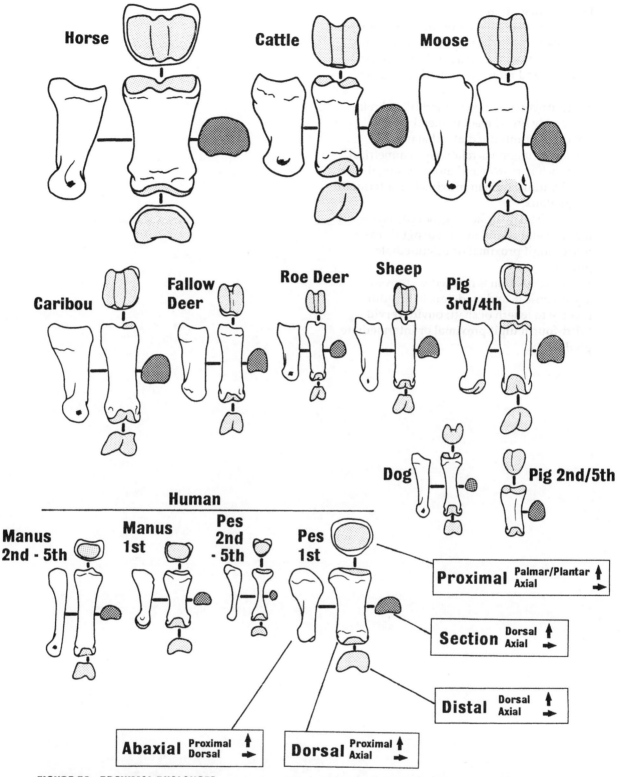

FIGURE 58. PROXIMAL PHALANGES

The proximal phalanges of quadrupeds are similar in the manus and the pes. Although a distinction can be made in the skeleton on one individual, this is not practicable in fragmentary material and this figure illustrates just "proximal phalanges" for these animals. In humans, however, a clear distinction can be made not only between manus and pes proximal phalanges, but also between the phalanges of the first digit and the other digits. Dorsal, proximal, distal and abaxial views, with a mid-shaft section, of horse, cattle, moose, caribou, fallow deer, roe deer, sheep and dog proximal phalanges. Similar views are shown of a pig third or fourth digit proximal phalanx, with dorsal and proximal views (and a section) for the much smaller pig second or fifth digit proximal phalanx. Dorsal, medial, proximal and distal views, with a mid-shaft section, for a left human manus proximal phalanx (generalised), manus first digit proximal phalanx, pes proximal phalanx (generalised) and pes first digit proximal phalanx. Approximately half life-size.

Distal articulation

• The distal articulation of the **intermediate phalanx** is generally narrower medial-lateral or axial-abaxial than that of the **proximal phalanx**, and more asymmetrical in most animals.

• The articulation is symmetrical and relatively very broad medial-lateral; **horse proximal** or **intermediate phalanx.**

• The articulation is relatively symmetrical but much narrower and more rectangular in distal outline; **human proximal** or **intermediate phalanx.**

• The articular surface is grooved, asymmetrical and relatively broad; **pig third-fourth digit proximal** or **intermediate phalanx.**

• The articulation is markedly grooved, asymmetrical, and moderate in breadth relative to length of shaft; **bovid, cervid third-fourth digit proximal** or **intermediate phalanx.**

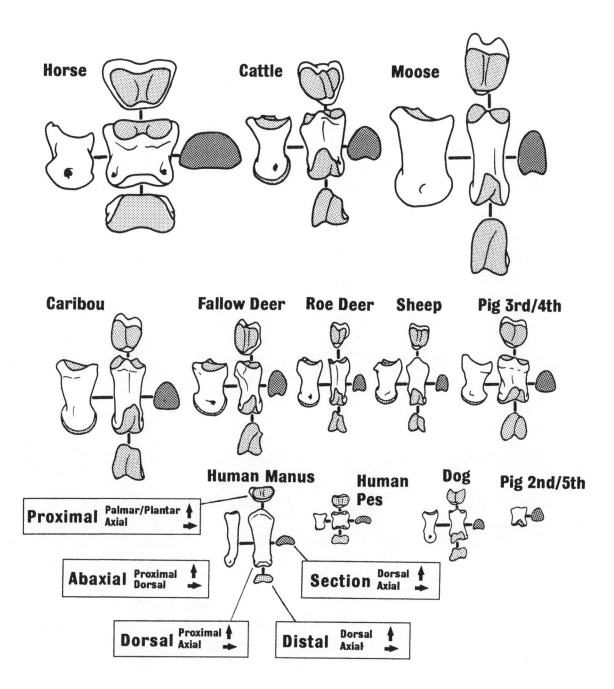

FIGURE 59. INTERMEDIATE PHALANGES

The intermediate phalanges of most mammals are difficult to distinguish between manus and pes, left and right. This figure therefore illustrates just generalised intermediate phalanges, except in the case of humans and pigs, where differences are readily apparent. Dorsal, abaxial, proximal and distal views, with a mid-shaft section, for the intermediate phalanx of horse, cattle, moose, caribou, fallow deer, roe, sheep and dog. Similar views of the third or fourth digit intermediate phalanx of pig, with a dorsal view and section of the pig second or fifth digit intermediate phalanx. Dorsal, medial, proximal and distal views, with a mid-shaft section of human left intermediate manus and pes phalanges. Approximately half life-size.

REFERENCES

Boessneck, J.
   1969   Osteological Differences between
   Sheep (*Ovis aries* Linné) and Goat (*Capra
   hircus* Linné). In Brothwell, D.R. &
   Higgs, E.S. (Eds), *Science in Archaeology*.
   2nd Edition. London: Thames & Hud-
   son. pp 331-358.
Boessneck, J., Müller, H.-H. & Teichert, M.
   1964   Osteologische
   Untersheidungsmerkmale zwischen
   Schaf (*Ovis aries* Linné) und Ziege (*Capra
   hircus* Linné). *Kühn Archiv* 78 (1/2). pp.
   1-129.
Clutton-Brock, J., Dennis-Bryan, K.,
   Armitage, P.L. & Jewell, P.A.
   1990   Osteology of the Soay Sheep. *Bulle-
   tin of the British Museum (Natural History)
   Zoology* 56(1) pp 1-56.
Corbet, G.B. & Hill, J.E.
   1986   *A World List of Mammalian Species*.
   2nd Edition. London: British Museum
   (Natural History).
Getty, R.
   1975   *Sisson and Grossman's The Anatomy of
   the Domestic Animals*. 5th Edition. 2
   Volumes. Philadelphia, London &
   Toronto: WB Saunders Co.
Hillson, S.W.
   1990   *Teeth*. Cambridge Manuals in
   Archaeology. Cambridge: Cambridge
   University Press.
Jayne, H.
   1898   *Mammalian Anatomy: a Preparation
   for Human and Comparative Anatomy. Part
   I: The Skeleton of the Cat*. London:
   Lippincott.
Kapandji, I.A.
   1974   *The Physiology of the Joints. Volume
   Three, The Trunk and the Vertebral Column*.
   2nd Edition. Edinburgh: Churchill
   Livingstone.
Lawlor, T.E.
   1979 *Handbook to the Orders and
   Families of Living Mammals*. 2nd
   Edition. Eureka: Mad River Press.
McCuaig Balkwill, D. & Cumbaa, S.
   1992 A Guide to the Identification of
   Postcranial Bones of *Bos Taurus* and
   Bison bison. *Syllogeus* No. 71. Ottawa :
   Canadian Museum of Nature

Olsen, S.J.
   1960   Post-Cranial Skeletal Characters of
   Bison and *Bos*. *Papers of the Peabody
   Museum of Archaeology and Ethnology,
   Harvard University* XXXV (4). Cambridge
   Massachusetts: Peabody Museum.
Pales, L. & Garcia, M.A.
   1981   *Atlas Ostéologique pour servir à
   l'identification des Mammifères*. Paris:
   Editions du Centre National de la Re-
   cherche Scientifique.
Payne, S.
   1969   A metrical distinction between
   sheep and goat. In Ucko, P.J. &
   Dimbleby, G.W. (Eds), *The Domestication
   and Exploitation of Plants and Animals*.
   London: Duckworth. pp 295-305.
Payne, S.
   1985   Morphological Distinctions between
   the Mandibular Teeth of Young Sheep,
   Ovis, and Goats, *Capra*. *Journal of Ar-
   chaeological Science* 12:139-147.
Prummel, W. & Frisch, H.-J.
   1986   A Guide for the Distinction of
   Species, Sex and Body Side in Bones and
   Sheep and Goat. *Journal of Archaeological
   Science* 13: 567-577.
Reynolds, S.
   1939   *A Monograph on the British Pleistocene
   Mammalia. Volume III; Part VI. The
   Bovidae*. London: Palaeontographical
   Society.
Schmid, E.
   1972   *Atlas of Animal Bones*. Amsterdam,
   London, New York: Elsevier.
Skerritt, G.C. & McLelland, J.
   1984   *An Introduction to the Functional
   Anatomy of the Limbs of the Domestic
   Animals*. Bristol: John Wright & Sons
   Ltd.
Taylor Page, F.J.
   1971   *Field Guide to British Deer*. 2nd
   Edition. Oxford & Edinburgh: Blackwell
   Scientific Publications.

INDEX

(faint, illegible text)

For Product Safety Concerns and Information please contact our EU
representative GPSR@taylorandfrancis.com Taylor & Francis Verlag GmbH,
Kaufingerstraße 24, 80331 München, Germany

Printed and bound by CPI Group (UK) Ltd, Croydon, CR0 4YY
01/05/2025
01858602-0001